Speaking Frankly About Conversation Intelligence

A Strategic Guide to Conversation Intelligence

By JC Quintana

CRG PRESS

Kennesaw, Georgia. United States

ISBN 978-0-9889145-2-0 (pbk.)

1st Edition June 2023

For

The incredible leadership and peers at congruentX!

You have taught me so much about the relevance of conversation intelligence and the innovative tools that make it so cool.

Avi

Chris

Chuck

Markus

Mike

Tap

Tasha

And the entire global team!

CONTENTS

PART 1:

It Is First About Conversation

Introduction

Because I know some of you reading this book are cautiously approaching the topic of conversation intelligence, let's start with a story.

The Story of Sweet Treats

Meet Sarah, the owner of a small family-owned bakery called Sweet Treats. The bakery had been in the family for generations. Sarah inherited it from her grandmother. Sweet Treats was known for its delectable pastries, cakes, and bread and had a loyal customer base. However, Sarah felt that the bakery could do better. She wanted to attract more customers and increase revenue.

One day, Sarah attended a business conference where she learned about conversation intelligence. She realized the key to growing her business was understanding her customers better and paying attention to what they said during their calls and interactions with her staff.

"They are constantly telling me what they like and don't like," Sarah thought. "I need to listen and understand what they want more of."

She knew her employees interacted with customers and vendors daily. Their conversations held valuable insights that could help her improve the business.

Sarah gathered her team and explained the concept of conversation intelligence.

"I learned that the many conversations we have with our customers are partners can be a gold mine of business intelligence," she told them. "We will pay more attention and incorporate some cool tools to help."

She encouraged them to take notes on customer conversations and share them with the rest of the gang. Sarah wanted to ensure they understood the value of dialogue and active listening. She trained them to recognize customer sentiment towards products and her brand and listen intently to customer insights. She started every morning before opening their doors with a friendly gathering.

"OK, team, what are customers saying, feeling, sharing? Let's talk about it!"

Then, one morning, Sarah announced that she discovered a more efficient way to gather insights from conversations with customers over the phone, email, website chat, and social media.

"Business is booming, team. It's time to get some help automating the process of gathering and analyzing customer conversations."

Improving the Business

Over time, the team at Sweet Treats became adept at gathering and analyzing conversation data. They discovered that their customers loved their pastries but wanted more variety. They also found that customers were willing to pay more for premium ingredients. The team at Sweet Treats used this information to develop new bakery goods and improve existing ones. They introduced new pastry flavors, experimented with premium ingredients, and even offered gluten-free options. They also used conversation data to improve their customer service.

As a result, Sweet Treats started attracting more customers. Word of mouth spread, and the bakery became known for its delicious and innovative products. The team at Sweet Treats also noticed that their existing customers were more loyal and spent more money at the bakery. They also received more positive reviews online, which helped build brand awareness.

Expanding the Use of Conversation Intelligence

Sarah realized she could apply conversation intelligence to other business areas. She gathered her team again.

"Conversation intelligence has done so much to help us understand our customers. Maybe it is also time to listen to our distributors, suppliers, and vendors too."

She encouraged her team to gather and analyze conversations with the people and companies that helped them daily. By understanding the needs and concerns of these stakeholders, Sweet Treats was able to negotiate better deals and partnerships. Sarah also used conversation intelligence to improve the operations of the business. She analyzed conversations between employees to identify areas where the team needed training. She also used conversation data to optimize the workflow of the bakery, reducing waste and increasing efficiency.

Perhaps the most significant impact of conversation intelligence was on customer loyalty and advocacy. Sweet Treats had always had a loyal customer base, but now they have customers who were faithful and advocates. Customers who loved the bakery's products and services were likelier to recommend it to their friends and family. And because Sweet Treats had taken the time to understand their customers' needs, they were able to provide a personalized experience that kept customers coming back. Paying attention also helped them improve their relationships with valuable partners, enabling them to keep the business going.

The Importance of Leadership

The story of Sweet Treats is a testament to the power of conversation intelligence in business. It shows that by listening to customers and gathering and analyzing conversation data, companies can succeed and leave a lasting impact on their customer communities. But more than that, it's a story about people who were passionate about what they did and worked together to achieve their goals. And that is the aspirational goal of every business, large or small.

Sweet Treats also demonstrates the importance of leadership in implementing conversation intelligence. Sarah was the driving force behind adopting conversation intelligence in her business. She recognized the potential of this approach, and she was willing to invest the time, resources, and effort needed to make it work.

Communication, Collaboration, and Innovation

As Sweet Treats continued to grow, Sarah realized that conversation intelligence was not just a tool but a mindset. It was about being curious, listening to customers, and being open to feedback. She encouraged her team to embrace this mindset and always look for ways to improve the business.

"Technology is awesome, but first and foremost, we must have a dialogue and pay attention mindset," Sarah always says.

Sweet Treats became more than just a bakery. Thanks to conversation intelligence, it became a community of people passionate about delicious food and excellent service. The impact of conversation intelligence on Sweet Treats was not just financial. It also had a profound effect on the team. They felt more connected to the business and each other. They were proud of their achievements and excited about the future.

The story of Sweet Treats is one of many similar stories that highlight the importance of customer-centricity in business. Sarah and her team recognized that their customers were the key to success. They prioritized understanding their needs and preferences and created products and services that delighted their customers and kept them returning.

The success of Sweet Treats also shows that conversation intelligence is not just a one-time effort. It's an ongoing process that requires continuous improvement and adaptation. As the business grew and changed, Sarah and her team had to adjust their approach to conversation intelligence. They had to incorporate new technologies, methodologies, and best practices to stay ahead of the competition.

Despite the challenges, Sarah and her team never lost sight of their goal. They remained committed to understanding their customers and providing them with the best possible experience. And that commitment paid off in the form of a thriving business and a legacy that inspires others.

The story of Sweet Treats is not just about technology and data. It's also about the power of conversations and collaboration. Sarah recognized that her employees interacted with customers daily, and she encouraged them to share their insights and ideas. By working together, the team at Sweet Treats could achieve more than anyone could on their own. But with the addition of technology that made understanding customers more effective, easier, and enjoyable, they were unstoppable.

Envisioning Your Conversation Intelligence Story

The Sweet Treats story is about the power of conversation. Let me say that again, the power of conversation. It's about how communication and collaboration can transform a business and create a legacy for generations. However, it is also about how a simple concept like conversation intelligence can lead to meaningful business success. Perhaps even inspiration.

I hope you noticed I did not share the Sweet Treats story from a technology perspective first. Telling the story of their needs and how they adopted a mindset that valued the insights stored within all their conversations, even before they implemented a tool to help them analyze them, is essential to the goal of this book.

This is a strategic guide to help you speak frankly about conversation intelligence within your organization. It is not a guide to implementing software that records, transcribes, and analyzes customer and prospect conversations (although it is part of our discussion). It is a guide to the perspectives, mindset, and strategy your company needs to benefit from such software. It is a book that will help you collaborate with the stakeholders that must discuss this topic strategically to make its implementation successful. If you get the strategy and mindset right, then the software can support the strategy.

For that reason, I want to emphasize that in this book, conversation intelligence refers to a company's strategy for improving conversation design and then to the technology that helps you gather and analyze it. Therefore, I dedicate the book's first part to the foundational aspects of conversation intelligence and then gradually introduce conversation intelligence technology planning and implementation in part two.

Using each chapter as a checklist of things to talk about will help you assess where you are in the journey:

- **Chapter 1: The Power of Words** explores the impact of language in business conversations, particularly jargon, and buzzwords that can be confusing and off-putting to customers. The chapter emphasizes the importance of clear and consistent messaging to build strong customer relationships and effective dialogue to achieve business outcomes. We also discuss the use of conversation intelligence as a tool to gain valuable insights into sales conversations and improve revenue operations. The chapter concludes with actionable steps for readers to improve their communication practices.

- **Chapter 2: The Conversation Intelligence Mindset** emphasizes the importance of adopting a conversation intelligence mindset when it comes to business conversations. The chapter highlights the power of words and how they can impact relationships, operations, and revenue. The conversation intelligence mindset values growth, dialogue, inclusivity, and lifelong learning. It emphasizes active listening, paying attention to nonverbal cues, and valuing diverse perspectives and experiences to create a more inclusive and respectful environment for communication regardless of technology. Adopting this mindset prepares your team for technology that supports behavior and a culture of attention.

- **Chapter 3: The Undeniable Business Outcomes** discusses the tangible business outcomes that conversation intelligence can provide for companies. It highlights how conversation intelligence can optimize sales, improve customer support, increase marketing effectiveness, and enhance operational efficiency. The chapter also provides real-world examples of companies across various industries that have successfully used conversation intelligence to gain a competitive edge. The chapter concludes by emphasizing the importance of identifying specific needs and goals, considering company culture, and evaluating available conversation intelligence tools before implementing a conversation intelligence strategy.

- **Chapter 4: The Infusion of Playbooks** discusses the importance of business methodologies and frameworks in helping professionals succeed. These frameworks offer a structured approach to various aspects of business operations, including sales, marketing, finance, and project management. The chapter further explores how conversation intelligence can be infused with these frameworks to provide more powerful tools for sales, service, marketing, and executive stakeholders. The chapter also highlights the importance of relationship development frameworks in building strong relationships with customers, suppliers, partners, employees, and other stakeholders. Finally, the chapter suggests using conversation intelligence to coach all company members to follow sales methodologies, giving cross-functional teams a competitive advantage in the marketplace.

- **Chapter 5: The Market Advantage** discusses the benefits of using conversation intelligence for market intelligence. Companies that use conversation intelligence to analyze customer feedback are more likely to achieve high customer satisfaction and retention levels. The chapter provides a list of considerations for market intelligence analysis, including understanding customer needs, identifying trends and patterns, improving customer satisfaction, enhancing customer engagement, identifying new sales opportunities, and improving employee training. The chapter also discusses the pros and cons of using conversation intelligence for market intelligence. It provides recommendations for success, such as defining clear objectives, choosing the right tool, training data analysts, ensuring data privacy and security, and focusing on actionable insights. The chapter concludes with some next steps, including continuously refining the approach to using conversation intelligence and integrating it with other data sources.

- **Chapter 6: The Influence of Sentiment** explores the importance of sentiment and polarity analysis in conversation intelligence. The chapter delves into the techniques used in conversation intelligence to determine sentiment and polarity in customer conversations. It covers keyword and linguistic analysis, machine learning, contextual analysis, entity recognition, emotion detection, tone analysis, social media analysis, and multi-lingual analysis definitions you should know. It also explains why sentiment and polarity analysis is important, including its ability to help companies understand customer attitudes and preferences, improve customer satisfaction and loyalty, identify potential leads, develop targeted marketing campaigns, improve product development, identify competitive advantages, and enhance brand reputation. The chapter also discusses the ROI of sentiment and polarity analysis.

- **Chapter 7: The Most Valid Concerns** discusses the privacy, accuracy, bias, legal, and transparency concerns of using conversation intelligence in businesses. The chapter emphasizes the importance of implementing appropriate privacy measures, obtaining informed consent from customers, involving diverse perspectives in developing and deploying conversation intelligence, complying with relevant regulations and privacy laws, and being transparent about data collection and usage practices. The chapter also provides recommendations for addressing these concerns and taking the next best steps, such as starting small, collaborating with all stakeholders, and continually monitoring and adjusting your conversation intelligence strategies.

- **Chapter 8: The Essential Planning** discusses the essential planning required before implementing conversation intelligence. The chapter emphasizes careful preparation, planning, and strategy to gather and analyze conversation data effectively. The chapter covers topics such as setting conversation intelligence goals, assessing current capabilities, and developing data collection, storage, management, quality, governance, and integration plans. The chapter highlights the importance of collaboration, benchmarking, and continuous monitoring and evaluation of conversation intelligence strategy. Additionally, the chapter emphasizes the need for a sound strategic business design to ensure the success of conversation intelligence.

- **Chapter 9: The Path to Insights** explores the benefits of creating visual dashboards and representations of conversation intelligence data and best practices for presenting this data to key stakeholders across a company. The chapter presents the case for using visual dashboards and presentations of conversation intelligence data to understand complex data sets better.

- **Chapter 10: The Future is Now** discusses possible paths for conversation intelligence in the immediate future. It addresses the extended use of AI and ML algorithms, natural language processing (NLP), voice recognition technologies, and their expected growth. This chapter also offers considerations for the impact of conversation intelligence that still requires additional research and innovation.

Congratulations on taking the first step toward unlocking the power of conversation intelligence within your organization. I encourage you to use this book as a conversation starter. Read each chapter and stop to engage pertinent stakeholders across your company in meaningful dialogue. By using the checklist provided in each chapter, you can assess where you are in your journey toward conversation intelligence success. Don't be afraid to start from the ground up and build your understanding of conversations' foundational words and concepts.

Embrace the technology available to you, but always remember that conversation intelligence is a concept before it is a tool. Use the following chapters to connect with real-world success stories, identify and address common concerns, and plan for the future of your business. Keep pushing forward and remember that the power of conversation intelligence is within your grasp.

Chapter 1:
The Power of Words

"Let's find some synergy...."

It means,

"working together to achieve a common goal."

But it feels more like,

"Let's throw some stuff at the wall and see what sticks."

We use the word "**leverage**" a lot, too. But when you dig deeper into the conversation, you find that for some people, "**leveraging**" was more about using power to get what they want.

Innovation conversations use the word "**streamline**" a lot these days. What fantastic word imagery for making things more effective, easy, and enjoyable. But when you listen carefully to the conversation, you realize that "streamlining" meant,

"We're going to cut corners and hope for the best."

Even the word "**innovative**," which you would think means "**new and exciting**," often translates to making a slightly different version of what already exists.

Of course, you want to give customers innovative and "**disruptive**" solutions because that is what your company does. You are "**changing the game**." But in the end, the conversation analysis yields a different message that sounds more like

"we're going to shake things up and hope for the best."

Your promise of "**scalable**" solutions rings like you are "**able to grow and expand**" value, but the conversation translates to

"we're going to make this thing bigger and hope for the best" instead.

But hey, you can always "**pivot**." You think it means "**changing direction**," but it often translates to

"we're going to abandon ship and hope our employees know how to swim."

And let's be honest, how often does "pivot" mean being nimble? It's more like,

"We're going to try something new and hope the customer doesn't notice."

Customer Buzzwords

Beyond the misuse of common business words and jargon, we also piece together phrases hoping to connect what we do with what it does for the customer. The "**customer-centric** " approach to business means "**putting the customer at the center of everything we do**." But listening more closely to your conversations with them sounds more like,

"We're going to do the bare minimum to keep our customers happy."

Similarly, we ask people to share challenges in the "**customer journey**," implying we are "**mapping out the customer experience from start to finish**." But how often does that happen? How often does the conversation go from exploring potential value in the journey to,

"We're going to pretend as if we care about your experience, but we're just trying to sell you more stuff."

Words Need Meaning

Words like "innovative," "disruptive," and "scalable" may sound impressive. However, without meaning and context, they become confusing and even off-putting if they don't translate into tangible customer benefits. Customers want to know how a business can help them solve their problems and meet their needs, not just how cool and cutting-edge they sound.

Ultimately, the words that matter most are the ones that create a clear and compelling value proposition for customers. These words help customers understand what they will get from a business and why it's worth their time and money. Communicating this message makes you more likely to win and keep customers over the long term.

It is crucial for customers to clearly and consistently understand the value that a company provides because value perceptions directly impact their decision to do business with that company. According to a survey conducted by HubSpot, 82% of consumers expect businesses to be consistent in their marketing and messaging across all channels. Customers want a clear and consistent message about a business's value in social media, email marketing, or face-to-face interactions and conversations. A study by McKinsey & Company also found that customers are willing to pay up to 20% more for a product or service if they perceive it as providing higher value than competitors.

Suppose your conversations with customers do not convince them that you have (or still have) the value they expect from you. In that case, your words are going to be meaningless. No matter how catchy or eloquent you are, every word in your customer conversations must align with their value and outcome expectations.

Relationships Need Words

The words you use in conversation are even more critical in relaying to customers that you have a relationship of mutual value and trust. While clear and consistent communication about a company's value is essential, it is equally important to establish a relationship of equity between businesses and customers. Both parties must feel they are getting what they need from the business relationship via open and transparent dialogue.

Building business relationships (and I emphasize "building" as an intentional and methodological construction process) must go beyond just discussing a company's value. It must focus on understanding the needs and goals of the customer, which requires active listening, empathy, and a willingness to adapt and evolve as the relationship progresses.

Conversations that do not show business value differ from ones that build a meaningful business relationship and helps enrich it. A conversation that doesn't show business value may be focused solely on the benefits of a product or service. In comparison, conversations that build a meaningful business relationship focus on understanding the customer's unique challenges and finding ways to solve them together.

Research has shown that building a solid equity relationship between businesses and their customers is crucial for long-term success. According to a study by Bain & Company, customers with a strong relationship with a company are more likely to stay loyal and refer others to that company. Customers with strong relationships with a company are up to 10 times more likely to refer others to it.

From Words to Dialogue

The average person speaks all these jargon, business, and relationship words at about 125 to 150 words per minute. We use hundreds or thousands of words to convey our thoughts and ideas. They all end up in conversations that vary depending on the topic of discussion, the level of formality, and the number of people involved.

The number of words that lead to meaning and understanding of their context can vary depending on the topic's complexity, the speakers' familiarity, and the level of detail required to convey the message. But the essential words in any conversation convey the main idea or message you intend to communicate. These "keywords" or "content words" are necessary for understanding the context and meaning of the conversation. We then include them in our dialogue with business stakeholders in the following ways, depending on what you do at your company:

- **Prospecting:** During the prospecting phase, dialogue with potential customers is essential to establish a connection and build trust. You do this through various channels like social media (LinkedIn is the social weapon of choice for me), email, and phone calls. By engaging in a dialogue with your potential customers, you can learn more about their needs and goals and tailor your approach accordingly.

- **Selling your value:** During the selling phase, dialogue with customers is crucial to communicate your value and address customer concerns or objections. Active listening and a willingness to adapt and evolve the messaging as the conversation progresses will be critical. Through customer dialogue, you build trust and demonstrate your expertise and commitment to understanding the outcomes customers hope to achieve through your products and services.

- **Customer acquisition:** Once a customer decides to do business with a company, dialogue is vital to ensure a smooth transition and establish expectations for the relationship. Acquisition conversations include discussing pricing, delivery timelines, and ongoing support. Engaging in a customer dialogue ensures meeting customer needs and building a solid foundation for the relationship.

- **Customer retention:** Dialogue with existing customers is important to maintain a strong and meaningful relationship over time. Regular check-ins, feedback sessions, and ongoing support will be important considerations. By engaging in a dialogue with customers, you demonstrate your commitment to their success and identify opportunities for improvement.

Your dialogue with customers is essential to achieving all your business outcomes and accelerating value for your business. It must happen throughout the various phases of the customer relationship to allow your company to establish a connection, build trust, and demonstrate your value and commitment to the customer experience.

Research has shown that dialogue is fundamental to tangible business outcomes and that neglecting it leads to severe repercussions:

- **Decreased Lead Generation:** Businesses may miss out on potential sales opportunities if they cannot communicate effectively with their target audience. According to a study by the Harvard Business Review, companies that engage in customer conversations to address inquiries within an hour were seven times more likely to qualify the lead than those that responded after an hour.

- **Poor Prospecting:** Prospecting approaches vary by salesperson style or the product or service you sell. However, prospecting is about the one-to-one conversation that increases the chances of showing the value of your offer (your value proposition). You will struggle to attract new customers if you cannot effectively communicate your value proposition. According to a survey by HubSpot, 40% of salespeople say that prospective conversations are the most challenging part of the sales process.

- **Decreased Customer Retention:** When you cannot effectively communicate with your customers and address their needs and concerns, customers may become dissatisfied and seek out competitors. According to a study by Bain & Company, even brief conversations can increase customer retention by 5%, increasing profits by 25% to 95%.

- **Negative Brand Image:** When customers feel you are not listening or addressing their concerns, they may share negative reviews or feedback on social media or other online platforms. Negative customer sentiments can damage your business's reputation and make it difficult to attract new customers. According to a survey by BrightLocal, 86% of consumers read reviews for local companies before making a purchase decision. Conversations allow you to counteract negative social sentiment.

Dialogue is essential to your business. It allows you to effectively exchange sentiment, perceptions, context, and ideas that clarify outcomes and the expectations that lead to them. Dialogue gives customers a voice and a sense of control and involvement in decision-making. It can improve the customer experience by creating a sense of community and shared values between the customer and the business.

Dialogue can increase customer retention by creating a sense of trust and commitment between the customer and the business. Customers who engaged in dialogue with a company were likelier to continue doing business with that business.

Incidentally, the Journal of Business and Psychology study found that dialogue can improve employee creativity by allowing employees to share ideas and perspectives. The study also found that dialogue improves employee commitment to the organization by creating a sense of shared purpose and values. So, don't forget your internal customers as you evaluate the impact of conversation intelligence practices and technology.

From Dialogue to Sales Outcomes

If you are a seller, I want to direct my attention to you immediately. You know that effective dialogue is crucial to your success. It's not just about pushing a product or service - it's about building a relationship with potential buyers. Engaging in meaningful dialogue builds rapport and establishes trust, leading to more sales.

Dialogue is a two-way conversation between you and the buyer. You can tailor your pitch or proposal to meet their needs by actively listening to their needs and concerns. You can uncover important information about their motivations, challenges, and priorities by asking open-ended questions. You can then use this information to make a compelling case for your product or service and overcome any objections they may have.

Effective dialogue can also help create a sense of urgency or scarcity, encouraging buyers to act before the opportunity is lost. By addressing their concerns and highlighting the benefits of your product or service, you can create a positive buying experience that leaves them confident in their decision.

If you are a sales manager, consider investing in training and development programs that promote practical dialogue skills that help you improve your sales performance. Effective dialogue is not just significant in an immediate sales context. Dialogue allows you and your team to build trust and rapport that opens the door to future opportunities, even when one does not exist.

From Dialogue to Conversation Intelligence

In addition to improving your dialogue and training your teams to be conscientious about what they say, there is tremendous value in the strategic and technology-assisted approach conversation intelligence offers. By leveraging conversation intelligence, you can gain valuable insights into the critical elements of sales conversations, including language, tone, and nonverbal cues.

Conversation intelligence is essential for sellers because it can help you understand the effectiveness of your sales approach. By analyzing your conversations with buyers, you can identify areas for improvement and develop strategies to address common objections or areas of confusion. This can help you improve your sales process's effectiveness and increase your chances of closing more sales.

But conversation intelligence isn't just about improving your sales approach - it's also about building stronger relationships with buyers. By analyzing the language, tone, and nonverbal cues used in your conversations, you can connect genuinely with the core needs of individuals and organizations. You can build trust and credibility with buyers, establish more robust relationships, and ultimately close more sales.

From Dialogue to Value Acceleration

Conversation intelligence helps sellers improve their sales performance and helps companies accelerate value for their customers. By analyzing sales conversations, you can gain valuable insights into your customers' needs, circumstances, and characteristics. You can also improve product development, marketing, and customer experience strategies.

Moreover, conversation intelligence can help companies improve the components of RevOps - revenue operations. RevOps is a holistic approach to revenue generation that involves aligning sales, marketing, and customer success teams. By leveraging conversation intelligence, companies can improve the alignment between these teams and ensure they all work towards a common goal - generating revenue.

For example, conversation intelligence can help marketing teams to develop more effective campaigns by identifying the language and messaging that resonates with customers. It can also help sales teams to identify the most effective sales tactics and objection-handling techniques for that message. And it can help customer success teams to align existing customers with that value and develop strategies to communicate sales enrichment messaging.

Sweet Treats Understood Dialogue

The success story of Sweet Treats is an excellent example of how a small family-owned bakery grew into a thriving business by understanding its customers better. The owner, Sarah, realized that the key to growing her business was listening intensely to what people were saying. She encouraged her team to share what people were saying because she realized that conversations are the diamond mine of every business. Sometimes conversations look more like clumps of meaningless coal. Still, if you mine them effectively, you will find, as Sarah did, that dialogue has much more value than you realize.

As I will repeatedly emphasize throughout the book, Sweet Treats' success was not solely about using technology. It was about understanding the customer better. Sarah's leadership and her team's dedication significantly impacted the business because she was open to listening and engaging in meaningful conversations as a human practice that leveraged technology to accelerate outcomes. The company that prioritizes understanding its customers' needs and preferences wins, no matter the size.

Conclusion & Next Steps

Words are even more powerful when you learn to use language to your advantage in business conversations. Business jargon can be confusing and off-putting to customers. Clear and consistent messaging is essential to building a solid relationship with them. Effective dialogue is even more critical to achieving your business outcomes. Conversation intelligence can help you accelerate the value of dialogue by providing insights into the sentiments and the context of words participants express.

1. Take note of your everyday business jargon and how customers perceive it.

2. Practice active listening and empathy in your customer conversations to better understand their needs and goals.

3. Invest in training and development programs promoting practical dialogue skills to improve sales performance.

4. Explore conversation intelligence tools to gain valuable insights into your sales conversations and improve revenue operations.

5. Engage in candid dialogue about the effectiveness of your communication practices as you contemplate using conversation intelligence to help you improve it.

Chapter 2:
The Conversation
Intelligence Mindset

Before pressing forward, please indulge me in summarizing some essential elements discussed in the previous chapter. You must pay attention to words' power when it comes to business. In the last chapter, we discussed how some words could be misused while others can transform business relationships, operations, and revenue. How you talk to your colleagues, clients, and customers can significantly impact your company's success. The words you choose and how you say them can influence people and companies in many ways.

We also discussed how dialogue helps you build relationships. There are more conversation intelligence and relationship expectations discussions in the chapters ahead. Remember that when you communicate effectively with others, you build trust and rapport. That trust leads to stronger relationships with clients and customers, more business, and increased revenue. If you take the time to listen to your clients and understand their needs, you can tailor your services to meet those needs.

On the other hand, the wrong words can damage relationships and even lead to lost business. If you're dismissive or rude to clients, they won't want to work with you again. Inappropriate language or offensive comments can damage your company's reputation and cause you to lose the trust of your customers. Maybe even lose your job.

Words can also influence a company's culture. Language leaders and employees can shape how people think and behave. Words impact decision-making. So, don't underestimate the power of words in business dialogue.

Aligning Good Practices with Good Tools

So, why the long discussion about good words and dialogue practices in a book written about conversation intelligence? Because conversation intelligence tools cannot compensate for bad behavior. Conversation intelligence will help make sense of your conversations with customers and peers by leveraging artificial intelligence to analyze them. That analysis enables you to improve communication skills and build better relationships. However, it is imperative not to see conversation intelligence solely as a tool but as a technology-assisted mindset and a way of thinking about dialogue and conversations. By understanding conversation intelligence as a mindset first, you can develop a deeper understanding of the underlying principles of effective communication that technology helps you discover.

At its core, conversation intelligence is about understanding the neuroscience of conversations and how they trigger physical and emotional changes in the brain. According to research by Judith E. Glaser, conversations can either open us up to healthy, trusting exchanges or shut us down and trigger fear and distrust. That is not something a tool alone regulates or influences. You must play an active role. By seeing conversation intelligence as a mindset, you can learn to recognize these patterns and take steps to create more positive and productive conversations.

Keep A "Growth" Mindset

One key aspect of the conversation intelligence mindset is the growth mindset, which is the belief that individual qualities and abilities can be developed through effort and practice. Research by Carol Dweck shows that individuals with a growth mindset are more likely to embrace challenges and see setbacks as necessary for learning and growth. By adopting a growth mindset towards conversation intelligence, you can approach conversations as opportunities for growth and learning rather than threats to avoid. You can see conversation intelligence's insights as beneficial and productive rather than "Big Brother" judging your performance.

Avoid Monologues

Another critical aspect of the conversation intelligence mindset is emphasizing dialogue rather than a monologue. Dialogue is a two-way, cooperative conversation where participants exchange information and build relationships. The moment the other person stops talking, it is no longer a dialogue; without dialogue you cannot assemble insights.

Avoiding monologue is important in conversation because it allows for a more engaging and fulfilling exchange of understanding and ideas. Monologues can lead to disengagement and boredom in listeners, while interactive conversations can increase engagement and satisfaction.

Psychologists have found that conversations that involve active listening and participation from both parties can lead to increased empathy and understanding between individuals. In a study by Weger, participants who engaged in active listening during a conversation reported higher levels of empathy and perceived understanding from their conversation partner.

By seeing conversation intelligence as a mindset that values dialogue over monologue, you can learn to listen more effectively and ask better questions. In turn, conversation intelligence will give you richer insights.

Watch Your Tone

Conversation intelligence is also about tone, body language, and facial expressions. According to a study by Mehrabian and Ferris, the tone of voice accounts for 38% of communication. In comparison, words only account for 7%, and nonverbal cues account for 55%. Therefore, the tone of voice used during a conversation can significantly impact how the listener perceives the message.

Research by Ambady and Rosenthal found that, as humans, we can accurately perceive emotional states based solely on vocal cues. The study suggests that the tone of voice used during a conversation can convey emotions and attitudes that may not be explicitly stated in our words.

Adopting a conversation intelligence mindset allows you to pay attention to these nonverbal cues and adjust the conversation. Conversation intelligence technology effectively identifies tone variations that denote positive and negative sentiment. However, your awareness of tone can help direct the conversation in the right direction.

Remember Context

There is much we can say about context and contextual influences impacting dialogue. The expression "context is everything" is even more accurate when referring to conversations. Where you are (physically), the nature of the interaction, and even the location (in person, via video, phone call) all influence the nature and quality of the conversation. Context is important in conversation because it can significantly impact how the listener interprets and understands messages.

People rely heavily on contextual information to understand the meaning of an utterance. Context provides important cues about the speaker's intentions, emotions, and attitudes, which can impact how the listener perceives the message.

The context in which a conversation occurs can also impact the power dynamics between individuals. In a study by Tannen, participants reported feeling more powerful and in control when they were in a familiar context compared to an unfamiliar context. This suggests that the context in which a conversation occurs can impact the power dynamics between individuals, affecting how messages are interpreted and understood. Context can also influence the social norms that govern conversation.

Conversation intelligence can give you valuable insights to help you navigate complex situations like conflicts and negotiations within various contexts. However, it is up to you to identify the context most suitable for conversation success.

Keep It Human

Conversation intelligence can also promote inclusivity and diversity in communication. By recognizing and valuing diverse perspectives and experiences, you can create a more inclusive and respectful environment for communication. Diverse teams are more innovative and productive than homogeneous teams. Adopting a conversation intelligence mindset that values diversity and inclusivity can create a culture of open communication where everyone's voice is heard and valued.

Conversational vs. Conversation Intelligence

As we conclude this chapter, let me acknowledge that it is easy to mistake the term "conversation" intelligence for "conversational" intelligence. The terms are often used interchangeably, but there is a subtle difference between the two.

As coined by Judith E. Glaser (whom I referenced earlier), conversational intelligence refers to the ability to connect with others and build trust through conversations. Conversational intelligence involves understanding the neurochemistry of conversations and how to create a safe space for open communication. It focuses on the quality of conversations and the ability to build relationships through effective communication.

On the other hand, conversation intelligence is a more general term that refers to the ability to analyze and understand conversations. Including analyzing the content of conversations, identifying patterns and trends, and using that information to improve communication and decision-making. Conversation intelligence focuses on the quantitative aspects of conversations and the ability to extract meaningful insights from them.

Over the years, advances in neuroscience have enabled researchers to study the brain activity of people while they are engaged in conversations. Today we leverage brain research to gain a deeper understanding of the cognitive and emotional processes that underlie communication.

Likewise, in cognitive psychology, several theories of attention explain how we selectively focus on specific aspects of our environment while ignoring others. One such theory is the "filter theory," which suggests filtering out irrelevant stimuli while attending to important information. Another theory is the "resource model," which proposes that attention is a limited resource that must be divided among different tasks.

Conversation intelligence, the study of communication patterns and their impact on outcomes, uses these theories to understand how people process and respond to information during conversations. For example, by understanding the filter theory, conversation intelligence can help you identify which aspects of a conversation you should focus on to achieve a desired outcome. Similarly, the resource model can help you understand how to allocate your attentional resources during a conversation to ensure that you can effectively process and respond to the presented information.

Overall, the theories of attention in cognitive psychology provide a valuable framework for understanding how individuals attend to and process information during conversations. So, as you learn more about the use of conversation intelligence and AI to give you insights, remember that this is all about paying attention to what matters most to the people in the conversation. It is about extracting insights (intelligence) from our conversations.

However, part of that gathered intelligence must include insights into the quality of your conversations and relationships. You could say that conversation intelligence can help you obtain greater conversational intelligence.

Sweet Treats Got the Point

The success story of Sweet Treats also had a lot to do with adopting the conversation intelligence mindset I introduce in this chapter. Sarah, the owner, recognized the power of words and how they can influence people and companies in many ways. She encouraged her team to gather and analyze customer conversations, which helped them develop new products, improve customer service, and build brand awareness. However, Sarah did not see conversation intelligence solely as a tool but as a technology-assisted mindset and a way of thinking about dialogue and conversations. By understanding conversation intelligence as a mindset first, she developed a deeper understanding of the underlying principles of effective communication. She used this knowledge to take her business in a new direction.

Sarah and her team adopted a growth mindset towards conversation intelligence, approaching conversations as opportunities for growth and learning rather than as threats to avoid. They emphasized dialogue over monologue, listening more effectively, asking better questions, and building stronger relationships through conversation. They paid attention to nonverbal cues such as tone, body language, and facial expressions, using them to adjust and fine-tune their unique approach. They also recognized and valued diverse perspectives and experiences, creating a culture of open communication where everyone's voice was heard and valued.

Sarah and her team created a lifelong commitment to learning and growth by seeing conversation intelligence as a mindset. They recognized that effective communication is a lifelong process that requires ongoing effort and practice. They continued to improve their skills and build stronger relationships throughout their business's growth.

Conclusion & Next Steps

Conversation intelligence is a powerful tool for improving communication skills and building relationships. However, it is essential to see conversation intelligence not just as a technology or a tool but also as a mindset and dialogue philosophy for conversational success.

- Understand and apply conversation intelligence as a mindset first and foremost, even when technology is not part of the discussion.

- Practice active listening techniques to improve your communication skills.

- Pay attention to nonverbal cues, such as tone, body language, and facial expressions, to enhance your communication skills.

- Embrace challenges and see effort as a necessary part of learning and growth.

- Recognize and value diverse perspectives and experiences to create a more inclusive and respectful environment for communication.

- Continue to improve your skills and build stronger relationships by adopting a conversation intelligence mindset that values growth, dialogue, inclusivity, and lifelong learning.

Chapter 3:
The Undeniable Business Outcomes

Now that we have established that the mindset and behavior of dialogue and conversation intelligence is the foundation for its success, let's talk about it as a technology-assisted resource. Let's also talk about it as an essential business advantage.

You probably know at this point that conversation intelligence can provide valuable insights into what's working and what's not by analyzing customer interactions. That alone will help you improve your sales team's performance. With the right tools and techniques, you can use conversation intelligence to coach your team to success and gain a competitive edge in your industry. But what about your company? What about accelerating business value for your business? Can conversation intelligence take your business strategy to the next level?

Conversation intelligence isn't just about improving your sales process. It can also help you identify trends and patterns in customer behavior, allowing you to make data-driven decisions about your business strategy. You can gain insights into your team's performance and adjust improve business operations by tracking metrics such as call duration, talk-to-listen ratio, and sentiment analysis.

Of course, implementing conversation intelligence isn't always easy. It requires the right tools, processes, and training to be effective. But with the right approach, conversation intelligence can be a powerful tool for driving growth across every area of your business. So, if you're ready to take your business to the next level, consider incorporating conversation intelligence tools into your strategy.

Functional Use Cases

Here are some use cases to get your brain percolating and to get you thinking about what conversation intelligence could do for your own customer segments and value proposition:

Sales Optimization

One company that has successfully used conversation intelligence to optimize its sales process is T-Mobile. T-Mobile uses conversation intelligence to analyze sales calls and provide insights into what works and what doesn't.

By analyzing sales calls, T-Mobile can identify which sales techniques are most effective and which are not and provide training to sales teams accordingly. As a result, T-Mobile has seen increases in customer satisfaction and retention rates rise over the years.

Customer Support

American Express uses AI to analyze customer conversations across all communication channels, including phone calls and emails. American Express can identify common customer issues by analyzing customer conversations and developing solutions. By identifying customers that frequently call about a specific topic, American Express can classify the root cause of the problem and develop a solution to address it proactively. American Express has reduced its call center volume and improved customer satisfaction rates via this approach.

Marketing Optimization

Marriott International uses conversation intelligence to optimize its marketing efforts. By analyzing customer conversations, Marriott International can identify which marketing messages resonate with customers and which do not and adjust their messaging accordingly. As a result, Marriott International has increased its customer engagement and loyalty.

Operational Efficiency

Delta Air Lines uses conversation intelligence to identify areas where they can streamline their processes and reduce costs. Customers call frequently about specific issues, and Delta can identify the root cause of the problems and develop solutions to address them proactively. As a result, Delta Air Lines has improved their customer satisfaction rates.

Industry Use Cases

Conversation intelligence is a game-changer for many industries (perhaps yours) because it gives you a competitive edge over other companies in your space. Companies from various industries get insights into customer behavior and preferences by using AI to analyze what customers say and the nature of conversations throughout their interactions.

This means they can tailor their messaging, products, and services to their target audience and understand their wants. That's a big deal in a world increasingly empowered by tools like AI. Let's look at some specific examples of how conversation intelligence is used in different industries:

- **Retail:** In the retail industry, conversation intelligence is used to gain insights into consumer needs, preferences, and behaviors. By analyzing customer conversations across the purchase cycle, retailers like Macy's can put more of what consumers need and want in front of them through their purchase channels.

- **Banking:** Capital One uses AI-powered tools in the finance industry to analyze customer conversations across all the channels customers use to manage their finances. Banks gain valuable insights into customer behavior and preferences this way, which allows them to tailor their card, loan, and online banking services to meet the specific needs of their diverse banking audiences.

- **Hospitality:** In the hospitality industry, analyzing customer conversations across all communication channels helps companies gain insights into the consistency of the guest experience across locations and interactions. Marriott International uses conversation intelligence to ensure guests have a consistently positive experience across all their sites.

- **Automotive:** Ford uses conversation intelligence in the automotive industry to improve their marketing efforts and drive customer engagement. Conversation intelligence allows them to pinpoint conversation insights about the many brands, models, and feature options of car buyers and act on them.

- **Education:** Arizona State University uses conversation intelligence in the education industry to improve student support services and drive better academic outcomes. By analyzing student conversations contextual to classroom and campus experiences, they gain insights that help them enhance academic performance and community safety.

- **Real Estate:** In the real estate industry, Zillow uses conversation intelligence to gain insights into developer, builder, and tenant behaviors and interactions. By analyzing conversations across email and unified communications, they improve living amenities and communities, ultimately improving customer engagement and satisfaction.

- **Energy:** In the energy industry, conversation intelligence helps companies like Duke Energy understand the sentiment on climate change and rising energy costs. Proactively communicating products and services that align with customer needs and complying with regulatory requirements helps them deliver products and services customers actually want.

- **Insurance:** Geico uses conversation intelligence in the insurance industry to gain insights into opportunities to cross-sell products they sell in partnership with other insurance vendors. Conversation intelligence allows Geico to tailor their offerings to the people who best align with its campaigns while complying with regulatory requirements.

- **Manufacturing:** Caterpillar uses conversation intelligence in the manufacturing industry to improve communication and collaboration among employees, suppliers, and customers. Analyzing customer conversations across all communication channels allows them to mobilize teams and communicate effectively during new product launches or recalls.

- **Food & Beverage:** In the food and beverage industry, Coca-Cola uses conversation intelligence to gain insights into consumer consumption preferences, accessibility, and demand. By analyzing customer conversations that reflect needs and trends across all communication channels, they can better orchestrate their communications and marketing efforts to meet customer needs.

- **Construction:** Home Depot uses conversation intelligence in the construction industry to proactively meet market and supply demands. They can improve their product offerings and customer satisfaction by analyzing customer conversations that reflect needs and trends, including contractor and homeowner needs.

- **Pharmaceuticals:** In the pharmaceutical industry, Pfizer uses conversation intelligence to gain insights into improving patient engagement and complying with regulatory requirements. By analyzing patient conversations across all communication channels, they can tailor their products and services to meet the needs of their customers better and ensure compliance with regulatory requirements.

I will stop here, but I encourage you to review additional industry-specific use cases for conversation intelligence in the reference section for this chapter. Among these industries are Media & Entertainment (Netflix), Agriculture (John Deere), Government (City of Boston), Non-Profit (American Cancer Society), and Sports (Golden State Warriors).

Maybe You Are a Sweet Treats Competitor

In the story of Sweet Treats, the company struggled to grow and compete in the highly competitive bakery industry. However, with the implementation of conversation intelligence, they were able to gain valuable insights into their customer behavior and make data-driven decisions about their business strategy.

Using conversation intelligence, Sweet Treats could track metrics such as call duration, talk-to-listen ratio, and sentiment analysis. These metrics allowed them to gain insights into their team's performance and make adjustments that significantly impacted their business operations. They could identify trends and patterns in customer behavior, allowing them to change their product offerings and marketing messaging.

For example, Sweet Treats identified that their customers were looking for more variety in their product offerings and were willing to pay a premium for organic and gluten-free options. With this information, they introduced new products and adjusted their pricing strategy to meet their customers' needs better. Conversation intelligence allowed Sweet Treats to improve customer support by identifying common customer issues and proactively developing solutions. It helped them reduce their call center volume and improve customer satisfaction rates.

Overall, the implementation of conversation intelligence was a game-changer for Sweet Treats, allowing them to gain a competitive edge in the bakery industry and drive growth across every area of their business.

Conclusion & Next Steps

These real-world examples demonstrate that conversation intelligence can be used effectively in many industries with diverse customer bases and missions. As businesses across industries continue to adopt conversation intelligence, we can expect to see even more innovative use cases in the future.

When considering the implementation of conversation intelligence to address specific needs for your industry, you should consider the following considerations:

- **Identify your company's and industry's specific needs and goals:** Before implementing conversation intelligence, it is important to identify the needs and goals unique to your customer segments. It will help to determine which conversation intelligence tools and techniques are most appropriate.

- **Consider the company culture:** Your company culture can influence the effectiveness of conversation intelligence. You should consider how conversation intelligence aligns with what makes your company "who" it is.

- **Evaluate the available conversation intelligence tools in your space:** Many conversation intelligence tools are available. However, not all are a good fit for some industries.

These are only a few of the many considerations we will discuss again. In the following few chapters, you will have the opportunity to evaluate needs further, address further concerns, and start building a conversation intelligence success plan.

Chapter 4:
The Infusion of Playbooks

Business methodologies and frameworks are essential tools that help you succeed by providing best practices and guidelines. These frameworks offer a structured approach to various aspects of business practice in almost every role. The methodologies and best practices you adopt have everything to do with accelerating success. If you know something works, backed by research, and tested by time, why not leverage it? Streamlining your processes improves efficiency and effectiveness and helps you achieve your goals faster.

One of the most significant benefits of business methodologies and frameworks is that they provide a structured approach to solving complex business problems. They help business professionals stay up to date with industry best practices and trends. Sales methodologies like Challenger Selling and Solution selling are popular with many companies. Digital marketing frameworks like the Inbound marketing methodology provide a structured approach to attracting, engaging, and delighting customers using digital channels. By following these frameworks, businesses can stay ahead of the curve and adapt to changing customer behaviors and preferences. Or the entire sales team can adopt a sales methodology that provides them with a common approach to understanding customer needs and closing deals effectively.

Framework – Infused Conversation Intelligence

So, what if you infuse a powerful tool like conversation intelligence with the guidance these methodologies provide sales, service, marketing, and executive stakeholders in their perspective roles? You could have an even more powerful tool by analyzing customer conversations against these frameworks' recommendations and validity.

Here are some ways conversation intelligence can help you follow these frameworks. I include some of the most popular here, but think about the frameworks and playbooks applicable to you and your company:

Sales & Key Account Management

Conversation intelligence can be very effective for the people responsible for winning and keeping customers in a sales or key account management capacity. It can help align with sales methodologies like Solution Selling, Challenger Sales, and other popular playbooks. By analyzing customer conversations, conversation intelligence can identify areas of improvement and provide recommendations for enhancing sales conversations. For instance, conversation intelligence can pinpoint the most effective messaging and objection-handling techniques, enabling sales reps to close deals more efficiently (key in Solution Selling).

With the help of conversation intelligence, key account management teams can also gain valuable insights into their customer relationships, identify areas for improvement, and develop strategies for strengthening those relationships. By analyzing customer conversations for customers' decision-making processes (key in Challenger Selling), KAMs can gain valuable insights into their customer relationships, identify areas for improvement, and develop strategies for strengthening those relationships. For example, conversation intelligence can help KAMs identify key stakeholders within an account, understand their needs and pain points, and tailor their communication to address those needs. It can also help KAMs identify potential upsell and cross-sell opportunities and potential risks to the account. With the help of conversation intelligence, KAMs can build stronger relationships with their key accounts and drive business growth. Conversation intelligence can help sales and customer success organizations successfully implement standard methodologies and playbooks like:

- Strategic Account Planning (SAP)

- Key Account Management (KAM)

- Customer Success Management (CSM)

- Relationship Management Framework (RMF)

- Value-Based Selling (VBS)

- Challenger Customer (CC)

- Miller Heiman Strategic Selling

- Spin Selling

- Target Account Selling (TAS), and others.

Service & Support

Conversation intelligence can help service teams identify areas of improvement in their customer interactions by leveraging time-tested methodologies and playbooks. Some popular methods and playbooks in this area include Customer Success Management (CSM), Relationship Management Framework (RMF), and Value-Based Selling (VBS). These methodologies can help you develop a customer-centric approach to service, cross-selling opportunities, and customer retention challenges.

By analyzing customer conversations for effectiveness and accessibility or likelihood to recommend you to others, you can increase success scores based on frameworks like the Customer Experience Index (CXi) and Net Promoter Sore (NPS).

Marketing

Marketing teams can also benefit by using conversation intelligence to improve the implementation of methodologies and frameworks. Some popular methodologies and playbooks in this area include Inbound Marketing, Account-Based Marketing (ABM), and Demand Generation. These methodologies can help you finetune your approach, and conversation intelligence can help ensure their successful implementation. ABM, for example, can identify key decision-makers within your target accounts and understand their needs and pain points during your conversations with them. Conversation intelligence can also help you determine how well people respond to inbound and outbound marketing campaigns (their sentiment and receptivity to offerings).

A study by the Aberdeen Group found that companies that use ABM with conversation intelligence achieve 200% higher revenue growth than those that don't. Companies that use Inbound Marketing with conversation intelligence achieve three times higher ROI than those that don't. By aligning your marketing efforts with specific methodologies and playbooks, you can ensure that you deliver the best possible marketing communication to your target audience, driving business growth and revenue.

Product/Services Improvement and Innovation

Let's not forget the people building products and services who can benefit from product improvement and innovation conversation insights. They often directly collaborate with sales, support, marketing teams, customers, and distributors. By analyzing customer conversations, conversation intelligence can provide valuable insights into customer needs, preferences, and pain points related to product development and customer adoption. Some popular playbooks in this area include Design Thinking, Lean Startup, and Agile.

Leveraging the Human-centered Design Thinking (of which I am a raving fan) mindset with conversation intelligence can result in a more targeted collection of customer pain points and an even more precise understanding of problems. Similarly, Lean Startup playbooks help you gather conversation intelligence data specific to the impact of products and services on your customer segments so you can be more proactive at the right time.

Using conversation intelligence with Agile playbooks, you can identify the most compelling features and functionalities and develop more efficient products that meet those needs. Conversation intelligence can identify customer decision-making processes and align user feedback to streamline Agile methodologies.

The McKinsey & Company study I referenced earlier also found that companies that use Agile with conversation intelligence achieve 60% higher success rates on their projects than those that don't. And companies that use Design Thinking with conversation intelligence achieve 75% higher success rates on their projects than those that don't.

And although we will not be spending a lot of time using conversation intelligence to build software products and applications, you should know that capability is already here. We are entering a new phase of collaboration in which we will use our customer conversations to create AI-generated applications in seconds.

Executive Decision-Making

Before wrapping up this section, I do not want to overlook that executive stakeholders across the business also use methodologies and playbooks to help measure business success and make operational business decisions. Some popular methodologies and playbooks in this area include Lean Six Sigma and (once again) Human-centered Design Thinking. These methodologies can help you develop a structured decision-making approach based on the insights you gain from discoverable customer and stakeholder dialogue.

Using conversation intelligence with Lean Six Sigma, you can identify areas of waste and inefficiency in your business processes and develop strategies for improving those processes. You can identify customer needs and preferences and build products and services that meet those needs more quickly and effectively. And by using conversation intelligence with Human-centered Design Thinking, you can gain a deeper understanding of general sentiment toward specific business units and the company at large.

Companies that use Lean Six Sigma with conversation intelligence achieve 50% higher productivity and 20% higher customer satisfaction than those that don't.

Your Own Secret Sauce

Of course, in addition to the wisdom playbooks bring to our organizations, there are the proven methods your company has gained over the years. I immediately think about companies like The Ritz-Carlton, famously known for the motto "Ladies and Gentlemen Serving Ladies and Gentlemen." The Ritz-Carlton's motto reflects the hotel's commitment to providing exceptional service to all guests, regardless of their status or background. It has become a hallmark of the Ritz-Carlton brand. It is a testament to the hotel's dedication to creating a luxurious and welcoming experience for all guests, starting with their unique approach to serving them. However, that mindset became a methodology that hotels worldwide use to improve the guest experience.

Your company's unique approach to business sets you apart from your competitors and drives your success. So, I do not want to discount that while practical and theoretical frameworks and playbooks are great tools, your company may have its own unique playbook based on your company's secret sauce. Using conversation intelligence to refine and improve that playbook over time is also a great strategy.

Relationship Development Frameworks

I want to give you another framework you may not have considered as you think about methodologies conversation intelligence can help you implement: Relationship Development and Expectation Management.

Making a case for building relationships rather than closing transactions with people is not difficult. As human beings, we have an innate inclination to build relationships with others. We are social creatures and need companionship, support, and connection with others. Building relationships helps us feel a sense of belonging and purpose and provides emotional and physical benefits.

Building relationships is also essential in business because it helps to establish trust, credibility, and loyalty between you and your customers, suppliers, partners, employees, and other stakeholders. Building strong relationships with customers makes you more likely to retain them and receive repeat business.

However, beyond these practical benefits, building business relationships can also help establish a positive reputation for a company. When a business is known for treating its customers, employees, and partners well, it is more likely to attract new customers and business opportunities. So, it should not be surprising that conversation intelligence is important to business relationship development no matter what you do, what industry you serve, or what methodology you use to run your business.

What would happen if you could evaluate relationship sentiment and strength rather than just general appetite for customer purchase interest level, loyalty, and advocacy? What if conversation intelligence could give you a clearer understanding of the nature and strength of the relationship based on the core elements of relationship expectation management?

Value Outcomes

Perception of value outcomes misalignment occurs when the buyer and seller have different expectations about the sale results. For example, the buyer may expect a quick return on their investment. At the same time, the seller may prioritize long-term customer lifetime value instead. Conversation intelligence can help identify these misalignments by analyzing the language the buyer and seller use during the conversation to communicate value outcome expectations. Sales methodologies like Value-based Selling and other playbooks discussed in this section also emphasize the importance of value outcomes. However, relationship psychology elements emphasize value in context with relationship type, longevity, and other essential factors you must consider.

Centricity & Culture

Cultural misalignment can also be a significant issue in managing business relationships. Different cultures have different communication styles and expectations. Individuals also have their own culture, which entails their values and personal beliefs. Conversation intelligence can help identify cultural misalignments by analyzing language patterns and identifying potential cultural biases.

Engagement

Level of engagement misalignments can occur when one party is more engaged in the conversation than the other. Lack of engagement can lead to misunderstandings and missed opportunities. Conversation intelligence tools greatly determine the talk-to-listen ratio and other measurements we associate with engagement. However, a well-designed conversation intelligence tool can give you an even deeper engagement analysis by identifying elements like interest and willingness to co-create value.

Accountability

Accountability misalignments can occur when one party expects the other to take responsibility for certain aspects of the sale. For example, the buyer may expect the seller to provide ongoing support after the sale. In contrast, the seller may not have communicated this expectation. Conversation intelligence can help align conversation data with other contractual obligation data like contracts and service level agreements (SLAs) to determine accountability. Imagine a conversation analysis of a failure on the part of your company to meet a contractual obligation that you can then mitigate proactively.

Knowledge

Knowledge misalignments can occur when one party has more knowledge about the product or service than the other. Or when the level of mastery one party expects is lower than expected. Or the buyer may expect the seller to have a deep understanding of their industry. We all bring knowledge expectations to our conversation that, when misaligned, can reduce credibility in our ability to deliver. Conversation intelligence can help identify knowledge misalignments by analyzing language patterns and identifying potential knowledge gaps.

Transparency & Trust

Transparency misalignments can occur when one party is not fully transparent about certain aspects of the conversation. For example, the seller may not disclose all the product or service costs, and the buyer may display a negative sentiment or response. Having the opportunity to identify this misalignment and then proactively respond with the information the buyer needs is a sure way to regain credibility and trust.

Speaking of trust, trust misalignments can occur when one party does not trust the other. Conversation intelligence can help identify trust misalignments by analyzing language patterns and identifying consistency and voice tone patterns that show distrust.

The Experience

Experience expectations misalignments can occur when one party expects "the experience" of a product or service to look, sound, or feel a specific way. Or the experience of working with a company or brand to make them feel a certain way. Conversation intelligence can help identify customer experience expectations misalignments by analyzing experiential and emotional elements of the conversation.

I use many selling interactions and conversations to illustrate the power of conversation intelligence. Perhaps because having healthy business relationship conversations is foundational to sellers' success. However, conversation intelligence helps you if you are a marketer, product manager, key account manager, customer success team member, or customer support professional.

Helping Us All Sell

In fact, I believe that conversation intelligence can benefit every company member. Conversation intelligence helps everyone improve profitability and productivity so we can win together.

Using conversation intelligence to collaborate in customer acquisition, retention, enrichment, and advocacy efforts can significantly impact your business's return on investment (ROI). By providing us with data-driven insights and recommendations to improve value conversations, conversation intelligence can help the entire organization sell products and services, improve overall performance, and increase revenue. I am not suggesting that we all wear the salesperson's name tag, but I am suggesting that we should all take a more active role in supporting the sales effort.

Companies should consider using conversation intelligence to coach all company members to follow sales methodologies that give cross-functional teams a competitive advantage in the marketplace. Some use cases illustrate the potential ROI of using conversation intelligence to help every organization member follow sales methodologies. This approach is, of course, required for sellers:

Improving Sales Rep Performance

By analyzing customer sales conversations, conversation intelligence can help sales reps identify areas where they can improve their sales techniques and follow their chosen sales methodology more effectively. However, for us who do not sell for a living, it is even more essential to recognize moments of truth when the customer communicates a need and an interest to buy. Being "opportunity conscious" helps sales reps close more deals and increase performance. It also allows the rest of us to recognize the emergence of opportunities to add value.

Reducing Sales Cycle Time

By providing sales reps with insights into how customers respond to their sales messages, conversation intelligence can help them tailor their approach to each customer's needs and reduce the time it takes to close deals. However, expediting sales and enabling sales teams to craft the right message is a skill we should all have, and conversation intelligence can help us gain it.

Improving Sales Coaching

By analyzing sales conversations, conversation intelligence can help sales managers identify coaching opportunities and provide targeted feedback to their sales reps. Conversation intelligence-assisted coaching helps sales reps improve their sales techniques and follow their chosen sales methodology more effectively. Coaching should be an essential part of presenting our value to support sales efforts and give customers what they need. Coaching is also how we learn to support internal stakeholders and partners.

But What About the Seller?

Despite the significant benefits we have listed, I do not want to overlook that conversation intelligence is far from the go-to tool for every sales professional. As conversation intelligence becomes more prevalent in the sales industry, sellers are starting to have mixed feelings about using it. Some sellers love it, while others hate it.

The Lovers

The sellers who love conversation intelligence can't get enough of it. They see it as a tool to help them improve their sales techniques and close more deals. They love the data-driven insights and recommendations that conversation intelligence provides and feel they have a secret weapon in their sales arsenal.

Jesse says,

"I used to feel like I was shooting in the dark when it came to my sales conversations. But with conversation

intelligence, I feel like I have a map and a compass. I know exactly where I'm going and how to get there."

Tasha says,

"I love how conversation intelligence helps me tailor my approach to each customer's needs. I can see what's working and what's not and adjust my sales pitch accordingly. It's like having a personal sales coach with me on every call."

The Haters

On the other hand, the sellers who hate conversation intelligence (which I know is a strong word to use) see it as a threat to their sales skills and a tool that undermines their expertise. They argue that conversation intelligence takes the human element out of sales and reduces them to robots following a script.

Tap says,

"I don't need a machine to tell me how to do my job. I've been in sales for 20 years and know what works and what doesn't. Conversation intelligence is just a fancy way of telling me I'm doing it wrong."

Stephanie says,

"I hate how conversation intelligence makes me feel like I'm constantly being watched. It's like having a boss looking over my shoulder, critiquing my every move. It's creepy, and it makes me uncomfortable."

The Middle Ground

Of course, not all sellers fall into the lovers or haters camp. Some sellers see conversation intelligence as a helpful tool that can help them improve their sales techniques. Still, they also recognize the importance of human connection and expertise in sales.

Chuck says,

"I think conversation intelligence can be valuable, but it's not a replacement for human skills and expertise."

While each seller's perspective on conversation intelligence is unique, each type has some general business pros and cons.

The Lovers:

Pros:

- Conversation intelligence can help sellers improve their sales techniques and close more deals.

- Data-driven insights and recommendations can give sellers a competitive advantage in the marketplace.

- Tailoring sales approaches to each customer's needs can help build stronger customer relationships and increase customer loyalty.

Cons:

- Over-reliance on conversation intelligence can lead to a lack of creativity and flexibility in sales approaches.

- Conversation intelligence may not be practical for all sales situations. It may not be able to capture nuances in customer interactions.

- Sellers may become too focused on data and lose sight of the human element of sales.

The Haters:

Pros:

- Sales expertise and intuition can be valuable in certain situations where conversation intelligence may not be effective.

- Sellers who rely on their own skills and expertise may be more adaptable and flexible in their sales approaches.

- Sellers uncomfortable with conversation intelligence may focus more on building personal customer relationships. Although, as you read in this chapter, conversation intelligence CAN help you build genuine customer relationships.

Cons:

- Sellers who reject conversation intelligence may miss valuable data-driven insights and recommendations.

- Over-reliance on intuition and expertise may lead to missed opportunities for improvement and growth.

- Sellers who reject conversation intelligence may be disadvantaged in a marketplace where competitors use data-driven insights to improve their sales performance.

The Middle Ground:

Pros:

- Sellers who use conversation intelligence as a tool while relying on their skills and expertise can have a well-rounded approach to sales.

- Balancing data-driven insights with human skills and intuition can lead to more effective sales approaches.

- Using conversation intelligence to identify areas for improvement can help sellers continuously improve their sales techniques.

Cons:

- Balancing data-driven insights with human skills and intuition can be challenging and may require constant adjustment.

- Sellers who rely on both conversation intelligence and their own skills and expertise may need to invest more time and effort into their sales approaches.

- Using conversation intelligence to identify areas for improvement may be overwhelming or demotivating for some sellers.

Sweet Treats and Playbooks

Sweet Treats rocket-fueled its business by leveraging playbooks to fuel its small business. They used various marketing strategies to attract customers, such as offering free samples, creating a loyalty program, and partnering with local businesses to cross-promote their products. They trained their sales staff to provide excellent customer service and upsell products to increase sales. They also used data analysis to track sales trends and adjust pricing and product offerings based on various sales frameworks.

And wise Sara used playbooks to implement efficient production methods and automate specific tasks, such as inventory management and order fulfillment, based on their success. They also prioritized food safety and quality control to ensure their products met high standards, always listening intently to conversations across the business.

But perhaps the most convincing use of conversation intelligence at Sweet Treats is how it helped them manage relationships and implement their secret sauce. You see, at Sweet Treats they have something called "The Sweet Sweet Way" that defines the types of relationships they want to build with customers. A training program even ensures new employees learn the Sweet Sweet Way culture. Conversation intelligence gave Sarah a powerful way to analyze the success of the practices that made her business successful.

Conclusion & Next Steps

Overall, each type of seller has its unique perspective on conversation intelligence, and the pros and cons will depend on the individual seller's approach to sales. However, balancing data-driven insights and human skills and expertise can lead to more effective sales approaches and better business outcomes. The best strategy combines proven sales methodologies, your secret sauce, AI-assisted process and behavioral improvement, and our customer love.

Identify the business methodologies and frameworks your company uses or could benefit from. Sales methodologies like Challenger Selling or Solution Selling, digital marketing frameworks like Inbound Marketing, and customer success management frameworks are great examples.

- Consider how conversation intelligence can help align your conversations with these methodologies and frameworks. Look for conversation intelligence tools to analyze customer conversations against these frameworks' guidance and validity and provide recommendations for enhancing conversations.

- Implement conversation intelligence across your sales, service, marketing, and product development teams to gain valuable insights into customer needs, preferences, and pain points related to product development and customer use.

- Train your sales reps, service, and marketing teams to use conversation intelligence to improve performance and follow their chosen methodologies more effectively. Encourage them to balance data-driven insights and human skills and expertise.

- Consider the pros and cons of using conversation intelligence. Work to find a middle ground that works for your company and your sales team. Remember that conversation intelligence is a tool that can help you improve your sales techniques. Still, it's not a replacement for human skills and expertise.

- Finally, continue to refine and improve your company's unique approach to business over time by using conversation intelligence to gather customer insights and feedback and leveraging proven methodologies and frameworks to drive business growth and revenue.

Chapter 5:
The Market Advantage

You may remember that in the early 2000s, Netflix was a DVD rental company that faced intense competition from Blockbuster, a well-established video rental chain. However, Netflix recognized the opportunity to use market intelligence to gain a competitive advantage.

Netflix began gathering data on customer behavior, including what movies they rented, how often they rented, and how long they kept them. This data allowed Netflix to create a recommendation algorithm that suggested movies to customers based on their viewing history. This approach was a breakthrough in the industry, providing a personalized experience for customers and helping them discover new movies they might not have otherwise considered.

Netflix also used market intelligence to identify a shift in customer behavior towards online streaming. In response, it invested heavily in developing its streaming platform, which allowed customers to watch movies and TV shows instantly over the Internet. This move proved a game-changer, allowing Netflix to offer a more convenient and cost-effective service than its competitors.

Today, Netflix is the world's leading streaming entertainment service, with over 200 million subscribers in more than 190 countries. The company's use of market intelligence to personalize the customer experience and anticipate shifts in customer behavior has been a key factor in its success. By leveraging data and insights to inform its strategy, Netflix has become a market leader in the highly competitive entertainment industry. It also created a generation of show bingers who continue to provide behavioral trend data that Netflix will use to develop the next version of its services.

Conversation intelligence practices and technology can also help you and your company understand your market. Studies conducted by the Harvard Business Review found that companies that use conversation intelligence to analyze customer feedback are more likely to achieve high customer satisfaction and retention levels. Forrester Research also found that companies that use conversation intelligence to analyze customer conversations are more likely to achieve higher alignment between the products and services they take to market and those that enjoy consistent customer advocacy.

Market Intelligence

Conversation intelligence is a great resource to help you identify trends and patterns in customer conversations and inform product development and market strategies. So, it is crucial to include market intelligence in this book as part of your conversation intelligence capabilities and planning strategy. There are other applicable market intelligence examples that I could fit into this chapter. However, here are the top considerations for how conversation intelligence can support your market intelligence analysis efforts:

- **Understand customer needs and preferences** to develop targeted solutions.

- **Identify trends and patterns** in customer behavior to develop new products and services that meet customer needs.

- **Improve customer satisfaction** in the use of products and services.

- **Enhance customer engagement** to identify the communication channels customers most respond to and adjust engagement preferences.

- **Identify new sales opportunities** for cross-selling to increase revenue and profitability.

- **Gain a better understanding of market segments** to develop targeted marketing campaigns and tailor products and services to specific customer segments.

- **Improve customer experience** response times, product quality, or service expectations.

- **Enhance brand reputation** by addressing negative feedback or improve social media presence that increases customer trust.

- **Better sales forecasting** to increase accuracy and make data-driven inventory management and pricing decisions.

- **Improve employee training** for handling customer complaints or providing product recommendations.

Market Intelligence Pros and Cons

While using conversation intelligence for market intelligence can provide various benefits, there are also some potential drawbacks. Let's summarize some of the benefits we have listed so far:

Pros:

- Conversation intelligence provides valuable insights from conversations about customer needs, preferences, and pain points across various channels. It gives you a deeper understanding of what your customers are looking for and how you can meet their needs.

- Conversation intelligence helps you stay ahead of the competition by identifying trends and patterns in customer behavior and preferences you can use to develop new products and services that your competitors may not have even fathomed.

- Conversation intelligence analyzes customer feedback and complaints you can use to improve your products and services and proactively help you improve them to retain more customers in the long run.

- Conversation intelligence helps you respond quickly to customer needs and preferences to improve the customer experience in real-time.

- Conversation intelligence provides data-driven insights and recommendations to make more informed decisions about product development, marketing campaigns, and pricing strategies, ultimately leading to increased revenue.

- Conversation intelligence is scalable. You can use it to analyze large volumes of customer conversations across multiple channels, providing a comprehensive view of customer behavior and preferences.

Cons:

While the previous benefits of conversation intelligence far outweigh the list I am about to share, it is important also to consider the drawbacks.

- Analyzing customer conversations can be time-consuming and requires significant resources, including data analysts and software tools. For smaller companies, this can be a substantial investment.

- Analyzing customer conversations raises privacy concerns as companies collect and analyze their customers' personal information. Companies need to be transparent with their customers about the collection and use of their data and ensure that they comply with all relevant privacy regulations. Please review the privacy and legal considerations I share with the pertinent stakeholders in this book.

- Conversation intelligence is limited to the channels where customer conversations occur, such as social media, email, and chat. Lacking access to communication channels may restrict you from having a complete picture of customer behavior and preferences. Some customers may not use these channels to communicate with companies. Sellers may also prefer to use their cell phones to communicate with customers, a channel often excluded by some companies in their strategy.

- Analyzing customer conversations can introduce bias if the data is not analyzed objectively. You must ensure your data analysts are trained to analyze data objectively and without bias.

- Analyzing customer conversations can be complex, requiring specialized skills and software tools. You must invest in data analysts and software tools to effectively analyze customer conversations.

- Analyzing customer conversations may not provide a complete picture of customer behavior and preferences. Conversations may lack context or not accurately represent the customer's true feelings. When analyzing customer conversations, you must consider this and ensure you are not making assumptions based on incomplete or inaccurate data.

Speaking frankly about conversation intelligence requires that you consider your capabilities honestly.

Sweet Treats Seized the Moment

Sarah is a smart cookie (punt intended) who seized the opportunity to become a leader in her market. Her conversation intelligence from multiple customer touchpoints gave her insights she used to develop targeted solutions and improve their products and services in ways her competitors did not understand. Sweet Treats analyzed customer conversations to better understand various market segments, including how specific similar businesses met pain points in her space, demographic, and regions where she expanded her business. They used this information to develop targeted marketing campaigns and tailor their products and services to specific customer segments.

Conclusion & Next Steps

You know that understanding your customers is vital to success. That's why conversation intelligence is such a powerful tool for market intelligence. You can gain valuable insights into customer needs, preferences, and behavior by analyzing customer conversations across various channels, products, and services.

- Start by defining clear objectives for conversation intelligence, such as identifying customer advocacy or brand loyalty.

- Choose the right conversation intelligence tool that meets your needs and provides the proper analysis and insights. If you are an international company or if you cater to the need of bilingual customers, choose a tool that supports multilingual conversations.

- Train your data analysts to effectively tune customer conversations to your market, product, or service. The words and terms you use in your industry may have a different interpretation from their everyday use. I could have a conversation about eating an Apple, watching a movie about an Amazon, seeing a Caterpillar, or applying for a Visa. All of these are also the names of popular brands.

- Ensure that you are collecting and analyzing customer conversations in a way compliant with data privacy and security regulations.

- Act on the insights provided by conversation intelligence to improve your products and services, increase customer satisfaction, and drive revenue growth.

- Encourage collaboration across departments, including sales, marketing, customer support, and product development, to ensure that insights from conversation intelligence are effectively integrated into business decisions.

- Monitor and measure the results of using conversation intelligence for market intelligence to identify areas where you need to improve your approach and refine your strategies.

- Continuously refine your approach to using conversation intelligence for market intelligence, including the tools and techniques used to analyze customer conversations.

- Ensure that the data used for conversation intelligence is high quality, including accurate and complete customer information.

- Develop a data governance strategy for conversation intelligence, including policies and procedures for collecting, storing, and analyzing customer conversations.

- Focus on actionable insights from conversation intelligence rather than just collecting data.

- Invest in training and development for data analysts and other employees using conversation intelligence for market intelligence.

- Integrate conversation intelligence with other data sources, such as CRM or website analytics, to provide a more comprehensive view of customer behavior and preferences.

Chapter 6:
The Influence of Sentiment

In previous chapters, I used the word "sentiment" to discuss the importance of being mindful of how customers feel about you, your brand, your company, or the nature of the conversation. In this chapter, I want to discuss how sentiment influences conversation and how conversation intelligence tools give us valuable insights about sentiment and polarity.

Have you ever been on a meeting or phone call and sensed something was wrong? Maybe the individual at the other end used a particular word with extra emphasis, or the tone of their voice did not match the expression on their face. We use emotional tone and attitude expression during dialogue to classify our conversations and determine the subsequent actions. Positive sentiment reinforces our message and encourages us to stay on the same conversation path. When it is negative, it can have the opposite effect. The sentiment and polarity of our conversation influence emotions and feelings.

Sentiment refers to a situation's emotional tone, which can

be positive, negative, or neutral. For example, suppose you're considering a job offer. In that case, the sentiment of the situation might be positive if the job is in your field of interest and offers a good salary and benefits. On the other hand, the sentiment might be negative if the job is in a location you dislike or if the salary and benefits aren't what you were hoping for.

Polarity, on the other hand, refers to a situation's degree of positivity or negativity. For example, a job offer might have a positive sentiment overall. Still, the polarity might be more negative if the salary and benefits are lower than expected.

When making decisions, we often weigh the sentiment and polarity of the situation against our own values and priorities to determine the best course of action. So, if you highly value work-life balance, a job offer with a positive sentiment but a negative polarity might not be the best fit for you.

Conversation intelligence helps us analyze conversations by using natural language processing (NLP) algorithms to identify the emotional tone and attitude expressed in a conversation. It helps us determine whether the sentiment is positive, negative, or neutral and the degree of polarity. For example, it can help you identify when the emotional tone shifts during a conversation about the value of your products or services. By analyzing these cues, conversation intelligence can help you understand the sentiment and polarity of the conversation and help you determine how to respond (or not).

Sentiment and Polarity

Conversation intelligence tools use several techniques to determine sentiment and polarity:

- **Keyword analysis** identifies words and phrases associated with positive or negative sentiment. For example, "amazing" or "great" are often associated with positive sentiments. Words like "frustrating" or "terrible" are often associated with negative sentiments.

- **Linguistic analysis** identifies linguistic features associated with positive or negative sentiment, such as adjectives or the tone of voice used in a conversation.

- **Machine learning** algorithms analyze large volumes of customer conversations and identify language patterns associated with positive or negative sentiment. These algorithms can also learn from new data over time, improving the accuracy of sentiment and polarity analysis.

- **Contextual analysis** understands the context of a conversation and identifies sentiment and polarity accordingly. For example, a conversation about a delayed flight may have a negative sentiment. Still, the sentiment may become positive if the customer service representative can resolve the issue quickly and efficiently.

- **Entity recognition** identifies specific entities, such as products or services, associated with positive or negative sentiment. Entity recognition helps companies identify areas to improve their offerings or marketing campaigns.

- **Emotion detection** identifies the emotions expressed in a conversation, such as happiness, anger, or frustration. Emotion detection provides you with a more nuanced understanding of customer sentiment and helps you tailor your responses.

- **Tone analysis** identifies the tone of a conversation, such as whether it is formal or informal, friendly, or hostile, or positive or negative.

Social Media Analysis

In addition, conversation intelligence tools can also analyze social media conversations to determine sentiment and polarity. It helps you understand how customers talk about your brand on social media and identify areas where you need to improve your social media strategy.

Multi-Lingual Analysis

Some conversation intelligence tools can analyze conversations in multiple languages, allowing you to gain insights into customer sentiment and polarity across different regions and markets.

There are many reasons why sentiment and polarity analysis are important. The most obvious is that it allows you to gain insights into an emotional expression that helps you measure a prospect's propensity to buy or an existing customer to stay. However, how sentiment influences the customer relationship (or your relationship with all stakeholders involved in the conversation) can take many shapes.

Sweet Treats Feels You

Sweet Treats applied sentiment analysis to determine the emotional tone of customer conversations and identify areas where they needed to improve the customer experience. By analyzing customer conversations, they were able to gain insights into customer attitudes and preferences, which helped them identify areas where they needed to improve the customer experience, such as product quality, customer service, or marketing campaigns. Additionally, sentiment analysis helped them identify potential leads and improve lead generation efforts by identifying positive sentiment and polarity.

Sweet Treats also used contextual analysis to understand the context of a conversation and identify sentiment and polarity accordingly. For example, if a customer complained about a delayed order, Sweet Treats would try to resolve the issue quickly and efficiently to turn the negative sentiment positive. Shifting sentiment helped them improve customer satisfaction and loyalty, increasing revenue and reducing customer churn.

Furthermore, Sweet Treats used tone analysis to identify the emotional tone of a conversation and tailor their responses accordingly. They also used social media analysis to determine sentiment and polarity, which helped them understand how customers were talking about their brand on social media and identify areas where they needed to improve their social media strategy.

Conclusion & Next Steps

Sentiment and polarity analysis is an important aspect of conversation intelligence that can help you understand how people feel. Beyond the sentiment functionality conversation intelligence offers, sentiment refers to a person's emotional state or mood. It encompasses a range of emotions that could be positive (such as happiness, joy, and love) or negative (such as sadness, anger, and fear).

Remember that:

- Sentiment can be influenced by various factors, including personal experiences, social and cultural context, and individual differences in personality and temperament.

- Sentiment is an essential concept in psychology because it helps us understand how people experience and express emotions, and how emotions can impact various aspects of our lives, such as our relationships, health, and overall well-being.

- Conversation intelligence tools help you identify and analyze all these factors to determine how to act or respond.

Chapter 7:
The Most Valid Concerns

One of the most common questions about conversation intelligence (once people get past the cool features) is usually about legal or privacy implications. It makes sense that you and your company want to understand the implications of analyzing conversation data, including private information and discussions with your customers and peers.

A report by the European Commission's High-Level Expert Group on AI titled "Ethics Guidelines for Trustworthy AI" outlines a framework for developing AI systems that are human-centric, transparent, and accountable. The report identifies legal and ethical concerns as a common reason for delaying or postponing the development and deployment of AI technologies, including conversation intelligence.

This chapter will explore these concerns in-depth and provide relevant information to help you make informed decisions about using conversation intelligence effectively and responsibly. By understanding the potential risks and problems associated with this tool, you can ensure that you're using it ethically and legally while still gaining valuable insights into your customers.

I encourage you to keep an open mind as we delve into this topic as you become better equipped to use conversation intelligence to improve your business without compromising your ethics or legal obligations.

Privacy Concerns

As you explore the potential of conversation intelligence, it's essential to consider the privacy concerns associated with its use. You may have already received a call from your chief privacy officer, who wants to discuss prioritizing customer privacy and using conversation intelligence ethically and responsibly.

Here are some critical points to consider in your conversation with your privacy officer or team. These can serve as a foundation for candid discussion (which is what this book is about):

- **Anonymizing data** protects customer privacy while allowing your business to gain valuable insights into customer needs and behaviors. By removing personally identifiable information from conversation data, you and your company can analyze the data without compromising customer privacy.

You can replace a customer's name, phone number, or email address with a unique identifier as a start to anonymize data while still getting valuable insight from their conversations.

- **Providing clear and transparent data collection** can help you build customer trust and improve their willingness to share personal information. By being transparent about how you will use conversation data, you can ensure the ethical use of conversation intelligence. Write and communicate clear privacy policies outlining how you will use customer data and who will access it. Include people from across your organization who interact with customers throughout their journey to get a realistic view of where and how customers share this data.

- **Allowing customers to opt out** of conversation data collection gives customers control over their data and helps build trust. It also increases customer willingness to share information. You can provide an option for customers to opt out of conversation data collection or to delete their data.

- **Obtaining customer consent** before collecting and analyzing their conversation data is crucial to ensuring the ethical use of conversation intelligence. By obtaining consent, businesses can ensure that customers know how their data will be used and can make informed decisions about sharing their personal information (including asking for consent before recording a conversation).

Accuracy Concerns

It is fair to be concerned about the potential for inaccurate insights and decisions based on complex and nuanced conversations. Conversation intelligence technology can sometimes interpret the nuances of human language, including sarcasm, irony, and cultural references inaccurately. Misinterpretations of the meaning of certain words or phrases could lead to inaccurate conversation analysis. Missing contextual information can also impact the interpretation of the conversation. In addition, conversation intelligence technology may not be able to account for nonverbal cues, such as facial expressions or tone of voice, which can impact the interpretation of the conversation.

Research has shown that using human analysts to review and verify conversation data can significantly improve accuracy. Adding a human element to the analysis process ensures that conversation data is interpreted correctly and that insights are accurate. Human analysts can review the output of conversation intelligence algorithms to check for errors, inconsistencies, or inaccuracies. Continually refining AI algorithms can also improve accuracy over time.

Keep in mind that accuracy concerns are not unique to conversation intelligence. Any data analysis tool can lead to inaccurate insights if not used correctly. However, by combining human analysts and continually refining AI algorithms, you can ensure that you are using conversation intelligence accurately and effectively.

Bias Concerns

In 2016 some academic research validated the concern that AI algorithms can perpetuate biases if incorrectly designed and monitored. That's because AI algorithms are only as unbiased as the data they are trained on. If the data used to prepare the algorithm is biased, then the algorithm will perpetuate those biases. If a conversation intelligence algorithm is trained on data biased against a particular race or gender, it may produce biased results.

Bias in the context of conversation intelligence refers to the presence of unfair or inaccurate assumptions or judgments in the algorithms used to analyze and interpret conversations. For example, an algorithm designed to identify "aggressive" language may be biased against certain cultures or dialects that use more direct language. Or if an algorithm is trained on customer service interactions with a predominantly male customer base, it may not be as effective at understanding and responding to female customers. Bias in conversation intelligence can lead to inaccurate or unfair assessments of individuals or groups, perpetuate stereotypes, and contribute to discrimination.

To address this concern, you must ensure that you design fair and unbiased algorithms. Use diverse data sets to train and test the algorithm for biases before deploying it. You can use data sets representative of the population you are analyzing or techniques such as counterfactual analysis to test for biases in the algorithm.

In the context of conversation intelligence, counterfactual analysis can be used to explore what might have happened during a conversation if different actions or words had been used. For example, if a salesperson is analyzing a call with a potential customer, they might use counterfactual analysis to consider what might have happened if they had used a different approach or asked different questions.

Continually monitoring and adjusting the algorithm can help eliminate biases over time. Keeping track of the algorithm's output and adjusting ensure that it produces fair and unbiased results must be part of your conversation intelligence processes.

It's also essential your company involves diverse perspectives in developing and deploying conversation intelligence. Involve people from different races, gender, and cultural backgrounds in the development and testing of algorithms.

By doing so, you can ensure that your algorithms are designed to be fair and unbiased for everyone.

Legal Concerns

Perhaps you're meeting with the chief privacy officer was followed by a call from your legal department concerned about the potential legal implications of conversation intelligence. Some of the industries we previously discussed will require it. Industries that handle sensitive personal information, such as healthcare and finance, often have difficulty implementing conversation intelligence due to legal concerns. These industries are subject to strict regulations and privacy laws, such as HIPAA and GLBA, which govern personal data collection, storage, and use. Misusing or mishandling this information can have severe legal and financial consequences.

To address these concerns, be sure to include the following considerations in your conversation intelligence legal discussions:

- First, of course, comply with all relevant regulations and privacy laws. Understand the requirements of these laws and take the necessary steps to comply, including implementing appropriate security measures to protect personal information. Encrypting conversation data to protect it from unauthorized access is a must.

- Second, consider using a secure and compliant conversation intelligence platform designed specifically for your industry. These platforms are often built with the necessary security and compliance features to ensure you use conversation intelligence safely and responsibly.

- Third, involve legal and compliance experts in developing and implementing your conversation intelligence strategy. Experts can help you ensure compliance with all relevant laws and regulations and that your plan is designed to protect the privacy and security of personal information. They can help you develop a privacy policy outlining how you will use conversation data and who will access it.

- Finally, be transparent with your customers about how you are using conversation intelligence and protecting their personal information.

Transparency Concerns

Some companies worry that if customers know you record calls or use conversation data in your analyses, they will hesitate to opt-in to calls or speak transparently. Your customers may be concerned their conversations are being analyzed without their knowledge or consent, which can lead to a loss of trust and loyalty.

We already proposed some of these recommendations for other concerns in this chapter, but they are worth repeating specifically about transparency:

- **Be upfront:** Make sure your legal team understands exactly how the conversation intelligence tool works, what data it collects, and how that data will be used. Be transparent about the benefits and potential risks of using the tool and ensure that your legal team has access to all relevant documentation and information.

- **Get consent:** Make sure you have the consent of all parties involved in any recorded conversations. This means informing them that the conversation will be recorded, what the recording will be used for, and giving them the option to opt-out if they don't want to be recorded. Additionally, make sure you have a clear policy in place for handling any requests for access to recorded conversations.

- **Communicate clearly:** Ensure that all employees are aware of the conversation intelligence tool and how it will be used. This includes providing clear guidelines on what types of conversations will be recorded, who will have access to the recordings, and how the recordings will be used.

- **Protect data:** Ensure that all data collected by the conversation intelligence tool is stored securely and in compliance with relevant data protection laws. This includes implementing appropriate access controls, encryption, and data retention policies.

Sweet Treats Was Thinking Ahead

As a business owner in the food industry, Sarah was careful to address the privacy concerns related to conversation intelligence by implementing appropriate privacy measures and obtaining informed consent from customers before collecting and analyzing their data. After all, many food businesses had gone bankrupt due to data breaches. She was careful to anonymize data to protect customer privacy while gaining valuable insights into customer needs and behaviors. Additionally, Sarah provided clear and transparent information about data collection and use to build customer trust and improve their willingness to share personal information. Sarah always allowed customers to opt out of conversation data collection to give them control over their personal data and help build trust.

To address accuracy concerns, Sarah had part-time (then full-time) human analysts review and verify conversation data to ensure the accurate interpretation of conversation data and insights. She continually refined AI algorithms to improve accuracy over time by analyzing the results of previous uses of conversation intelligence and identifying areas for improvement.

Sarah expanded her business to offer international variations of her food to various demographics. To address bias concerns, she ensured that her algorithms were designed to be fair and unbiased by using diverse data sets to train the algorithm and test it for biases before deploying it. She wanted her insights to reflect accurate perceptions of her diverse, multinational audience.

Sarah always hoped Sweet Treats would become a national franchise, so to address legal concerns related to franchising her company, Sarah ensured that she was complying with all relevant regulations and privacy laws, such as the Franchise Rule and the General Data Protection Regulation (GDPR). She considered using a secure and compliant conversation intelligence platform designed specifically for the food industry. Sarah involved legal and compliance experts in developing and implementing her conversation intelligence strategy to ensure compliance and protect the privacy and security of personal information.

Our friend Sarah coverer her bases across each concern related to conversation intelligence by taking appropriate measures to ensure ethical and responsible use of the tool while still gaining valuable insights into her customers, which was crucial in her decision to franchise her company.

Conclusion & Next Steps

While the legal and ethical aspects of conversation intelligence must be part of your discussions with business and IT stakeholders, it does not have to be a showstopper. Keep in mind the advice provided in this chapter and consider the following next steps:

- **You don't have to boil the ocean:** Maybe your organization is unprepared for conversation intelligence. Start small and experiment with conversation intelligence on a small scale before implementing it across the company. Starting small will allow you to test the technology and identify potential issues or concerns before scaling up.

- **Collaborate on it:** Involve the pertinent stakeholders in the conversation intelligence implementation process. Gather applicable employees, customers, and legal and compliance experts. By involving the right people early in the process, you get everyone on the same page and address their concerns.

- **Follow the rules:** Ensure you comply with all relevant regulations and privacy laws. Understand HIPAA and GLBA requirements (and other applicable laws) and take the necessary steps to comply, such as implementing appropriate security measures to protect personal information. If you are in California, the California Consumer Privacy Act (CCPA) may apply. General Data Protection Regulation (GDPR) is applicable to many of our European friends. Personal Information Protection and Electronic Documents Act (PIPEDA) for Canada. There are other regulations and laws not included in this book.

- **Be Transparent:** With your employees, customers, and internal stakeholders about how you plan to use conversation intelligence and how you will protect their personal information. It is the most crucial step to build trust and confidence in conversation intelligence as an ethical and responsible tool for gathering insights.

- **Track progress:** If you start small, continually monitor, and adjust your conversation intelligence strategy to ensure that it produces fair and unbiased results. Keep track of the algorithm's output and adjust ensure that it has fair and unbiased results.

Now that we've addressed some of the conversation intelligence concerns, it's time to move on to the next chapter and start thinking about strategy and planning.

PART 2:

Your Conversation Intelligence Strategy

Chapter 8:
The Essential Planning

I hope the previous chapters laid a foundation for conversation intelligence concepts and assessment. Understanding the influence of words and the power of dialogue to demonstrate your value is the heart of conversation intelligence, the Transfer Architecture this AI technology came from, and the theories of attention the technology is based upon. Understanding the psychology, principles, technology, and concerns related to conversation intelligence will prepare you for this chapter.

Conversation intelligence has gained traction in recent years, with more and more companies adopting it to improve customer interactions and sales processes. The positive reception of conversation intelligence has been largely due to the many benefits listed in this book. If your company plans to compete primarily based on customer experience in the near future (which is 89% of companies, according to Gartner), conversation intelligence can help. We know that companies that use conversation intelligence to analyze their sales calls see a 28% increase in win rates and a 15% increase in revenue (InsideSales.com). So, the case for conversation intelligence is strong.

However, implementing conversation intelligence requires careful preparation, planning, and strategy to ensure that you can gather and analyze conversation data effectively. This chapter will discuss the steps you should take before implementing conversation intelligence.

By taking these steps, you can ensure you're ready to implement conversation intelligence successfully. You can identify your goals and objectives, assess your current capabilities, develop a data collection and analysis plan, and create an action plan based on the insights you gain from your conversation data.

Remember that every business is unique, so it's essential to tailor your approach to conversation intelligence to your specific needs and goals. By doing so, you can maximize the benefits of this tool and achieve your desired outcomes. Don't forget, however, that, like most technology, strategy guides success. Not the other way around. In my other book in this series, I say that technology like CRM cannot be successful without a "customer relationship management" strategy because THAT is what CRM is. I echo this sentiment by saying that a conversation intelligence tool without a sound strategic business design will fail.

I encourage you to leverage methodologies like Human-Centered Design Thinking and the tools it provides to collaborate with key stakeholders at your company and answer the following questions:

Are You Ready to Set Conversation Intelligence Goals?

Remember, being prepared and strategic in your approach to conversation intelligence is critical for success. Identifying actionable steps will ensure you're ready to implement conversation intelligence effectively and achieve your desired outcomes. Consider what you hope to accomplish with conversation intelligence to define your goals and objectives. Do you want to improve customer satisfaction, increase revenue, reduce churn, or do all the above?

Although this is not a project management book, SMART objectives are great for conversation intelligence planning. SMART objectives should be specific, measurable, achievable, relevant, and time bound. I also suggest following the SMART objectives approach because throughout the book I will continue to point out specific areas of planning, implementation, and benchmarking that require this approach to validate conversation intelligence success. It is not enough to say,

"We will measure the success of 100 calls next month."

You must be more intentional than that with measurements that look more like,

- **Sales:** Increase the number of sales calls that result in a one-on-one conversation with the decision-maker by 20% within the next quarter.

- **Marketing:** Improve customer engagement by 15% by analyzing customer feedback and identifying areas to improve conversations.

- **Customer Support:** Reduce customer wait times by 30% by analyzing call data and identifying areas where we can streamline conversations.

- **Product Development:** Improve product adoption by 10% by analyzing customer feedback and identifying areas where we can tailor conversations to customer needs.

- **Human Resources:** Increase employee satisfaction by 15% by analyzing employee feedback and identifying areas where we can improve psychological safety.

As for recommended percentages, setting challenging but achievable goals is essential. A good rule of thumb is to aim for a 10-20% improvement in each area, which can vary depending on your company's needs and current performance levels.

Leveraging Human-centered Design Thinking

Understanding where to start will require collaboration with your internal stakeholders. Here, again, Human-centered Design Thinking workshops create an effective forum for conversation intelligence planning.

Human-centered design thinking is a problem-solving approach that focuses on understanding the needs and behaviors of the people using a product or service. It involves empathizing with users, defining the problem, ideating solutions, prototyping, and testing those solutions. This approach can be beneficial in planning a conversation intelligence strategy because it ensures the strategy is designed with the end users in mind.

- **Sales Stakeholders:** Talk to your sales team to understand their current process and identify areas where you can integrate conversation intelligence. Work with them to create a plan for analyzing successful calls and identifying key conversation points.

- **Marketing Stakeholders:** Meet with your marketing team to discuss customer feedback and identify areas where you can improve conversations. Work with them to create messaging that addresses customer pain points and encourages engagement.

- **Customer Support & Experience Stakeholders:** Collaborate with your customer support team to understand their current process and identify areas where you can streamline conversations. Work with them to create a plan for reducing customer wait times, improving overall satisfaction, and finding a point of friction in the customer journey.

- **Product Development Stakeholders:** Meet with your product development team to design ways conversation intelligence can discover product adoption and usage-related challenges.

- **Human Resources Stakeholders:** Talk to your HR team to understand employee feedback and identify areas where you can improve conversations. Work with them to create a plan for increasing employee satisfaction and engagement.

What Benchmarks Will Fuel Value Acceleration?

Benchmarking is important in conversation intelligence because it allows you to compare your performance to industry standards and identify areas where you can improve. By benchmarking your conversations against your competitors or peers, you can gain valuable insights into what's working and what's not and make data-driven decisions to improve your performance.

Benchmarks are critical to conversation intelligence planning and measurement because:

- **They help you Identify strengths and weaknesses:** By benchmarking your conversations, you can identify areas where you're performing well and areas where you need to improve. Benchmarking helps you focus on the areas that will most impact your performance.

- **They help you set aspirational goals:** Benchmarking can help you set realistic goals based on industry standards for your conversations. Benchmarks allow you to stay focused and motivated as you work towards improving your performance based on something sensible and attainable.

- **They help you measure progress:** By benchmarking your conversations over time, you can track your progress and see how your performance compares to industry standards. Benchmarks help you stay on track and adjust as needed.

- **They help you stay competitive:** Benchmarking can help you stay competitive by identifying best practices and improvement areas. It can help you stay ahead of the competition and provide better customer experiences.

Here are some examples of the previous points with recommended or suggested benchmarks specific to some of the organizations most likely to benefit from conversation intelligence:

- **Sales:** Industry benchmarks suggest that successful sales calls should have a close rate of 30-40% and an average call length of 20-30 minutes (may vary depending on the industry and type of product or service being sold).

- **Marketing:** Industry benchmarks suggest a 20-30% email open rate and a reasonable 2-3% click-through rate (may vary depending on the industry and specific campaign).

- **Customer Support:** Industry benchmarks suggest that the average customer wait time should be less than 2 minutes (some companies may strive for even shorter wait times to improve customer satisfaction).

- **Product Development:** Industry benchmarks suggest an average 30-40% product adoption rate (may vary depending on the type of product and the target market).

- **Human Resources:** Industry benchmarks suggest a reasonable employee satisfaction rate of 80-90% (some companies may aim for even higher rates depending on the industry and specific company culture).

These benchmarks can be helpful as a starting point for evaluating performance. Still, it's essential to remember that they may not be applicable or achievable for all companies or industries.

Additional Planning Discussions

Conversation intelligence goals and success measurements like benchmarks will help you begin with the end in mind. As management guru Peter Drucker said,

"What gets measured gets managed."

However, you need to consider several other important planning elements in your evaluation and planning of conversation intelligence. I will leave you with this list and encourage you to engage the pertinent stakeholders who can address these areas before you move forward with the adoption of conversation intelligence:

- **TECHNOLOGY:** Do you have the technology to gather and analyze conversation data effectively (tools for data collection, storage, and analysis)? To ensure that your business is equipped to handle conversation intelligence, you should evaluate your current technology and identify any gaps that may need to be addressed.

127

- **TOOLS:** Besides conversation intelligence technology itself, you should consider the tools you have to analyze conversation data, such as sentiment analysis, natural language processing, or other analytics tools. Evaluate your current tools and identify any gaps.

- **PERSONNEL:** Do you have the personnel to gather and analyze conversation data effectively? Include data analysts, data scientists, or other personnel with expertise in conversation intelligence. Evaluate the personnel you currently have in place and identify any gaps.

- **SKILLS:** Consider the skills and training needed to use conversation intelligence effectively across your organization. For some people using conversation intelligence will appear seamless. Anyone else analysand data and creating "prompts" to guide data discovery and gathering will probably need training (I included a quick definition of conversation intelligence prompts at the end of this chapter). Include training on data analysis tools, natural language processing, or other skills related to conversation intelligence. Evaluate the skills and training of your personnel and identify any gaps.

Data-Specific Considerations

- **TYPES OF CONVERSATION DATA:** What types of conversation data are most relevant to your goals and objectives? Include data from customer interactions, social media, surveys, and other sources.

- **DATA COLLECTION METHODS:** What are the most effective methods for collecting conversation data? This could include using tools for data collection, such as chatbots, social media monitoring tools, or surveys. We also discussed sources like emails and calls throughout the book.

- **DATA STORAGE AND MANAGEMENT:** Where will the conversation data be stored, and how will it be managed? Have a plan for storing and managing conversation data, including security and privacy considerations.

- **DATA QUALITY MANAGEMENT:** How will you ensure the collected conversation data is accurate, complete, and consistent? Establish processes for data quality management, including data cleaning and validation.

- **DATA GOVERNANCE:** What policies and procedures will you have to manage the collection, storage, and use of conversation data? Establish clear policies and procedures for data governance, including data security, privacy, and retention.

- **DATA ANALYSIS TECHNIQUES:** What techniques will you use to analyze the conversation data? Consider sentiment analysis, natural language processing, and text analytics. Businesses should identify the most effective methods for analyzing the collected conversation data.

- **DATA VISUALIZATION:** How will the conversation data be presented in a visual format? Include charts, graphs, and other visualizations that make understanding and interpreting the data more accessible. Businesses should identify the most effective visualizations for presenting the conversation data. There is more about conversation intelligence insights visualization later in the book.

- **DATA INTEGRATION:** How will the conversation data be integrated with other data sources to gain a more comprehensive view of customer behavior and preferences? Businesses should identify the most effective methods for incorporating conversation data with other data sources.

- **DATA GOVERNANCE:** What policies and procedures will the business have to manage the analysis and use of conversation data? Your company probably has clear policies and procedures for data governance, including data security, privacy, and retention, that will include conversation intelligence data.

Sarah's Grand Plan

You know by now that Sarah does nothing without a plan. She did her research and engaged the right experts in her future planning from the start. Sarah planned for Sweet Treats to become a national company, so she carefully identified her conversation intelligence goals and objectives. She set specific, measurable, achievable, relevant, and time-bound (SMART) objectives, such as increasing sales and improving customer satisfaction by exact percentages. Sarah collaborated with key stakeholders across her team and then her more significant business to assess their current capabilities and identify any gaps that may need to be addressed.

She developed a data collection plan that outlined how to gather conversation data, what types of data to collect, and how often to collect it. Sarah established clear policies and procedures for data governance, including data security, privacy, and retention. She identified the most effective techniques for analyzing and visualizing conversation data and integrating it with other data sources, like her Customer Relationship Management (CRM) system. Sarah also benchmarked Sweet Treats conversations against industry standards to identify improvement areas.

Her success came from setting aspirational goals based on industry benchmarks and continuously monitoring and evaluating her conversation intelligence strategy to ensure it was aligned with her business goals and objectives. Overall, Sarah was strategic and intentional in her approach to conversation intelligence to help Sweet Treats become a successful national company. That made all the difference from the start.

Conclusion & Next Steps

Implementing conversation intelligence requires careful planning and preparation. Considering the many factors discussed in this chapter and developing a comprehensive plan, you can make the most of conversation intelligence and use it to drive business success.

Planning and collaboration will help you make data-driven decisions to improve customer satisfaction, increase sales, and achieve other business goals that accelerate value for your business in every aspect of customer acquisition, retention, and enrichment of your people, products, and services:

- Identify your conversation intelligence goals and objectives, and develop specific, measurable, achievable, relevant, and time-bound (SMART) objectives.

- Collaborate with key stakeholders across your business to assess your current capabilities and identify any gaps that may need to be addressed.

- Develop a data collection plan that outlines how you will gather conversation data, what types of data you will collect, and how often you will collect it.

- Establish clear policies and procedures for data governance, including data security, privacy, and retention.

- Identify the most effective techniques for analyzing and visualizing conversation data and integrating it with other data sources.

- Consider hiring personnel with expertise in conversation intelligence or provide training to your existing team members.

- Benchmark your conversations against industry standards to identify areas where you can improve.

- Stay focused and motivated by setting aspirational goals based on industry benchmarks.

- Continuously monitor and evaluate your conversation intelligence strategy to ensure it aligns with your business goals and objectives.

- Finally, remember that conversation intelligence is a tool that requires a sound strategic business design to be successful.

A Quick Word About Prompts

In conversation intelligence, a prompt is a keyword or phrase that triggers a conversation intelligence tool to start recording or analyzing a conversation. You can use prompts to identify specific topics or behaviors that interest you, such as mentions of a particular product or service or using a specific language or tone.

For example, if you're a sales manager, you might set up a prompt to identify when a sales rep mentions a competitor's product so that you can analyze how the rep responds to that mention. Or suppose if you are a customer service manager, you might set up a prompt to identify when a customer expresses frustration so that you can evaluate how well the customer service agent handles that situation.

You can set up prompts manually or use conversation intelligence tools that automatically generate prompts based on natural language processing and machine learning. Once a prompt is triggered, the conversation intelligence tool can start recording or analyzing the conversation, giving you valuable insights into customer interactions and employee performance. By using prompts, you can focus on the topics or behaviors most important to you and use that information to improve your customer interactions and sales processes.

How well you write or "design" a prompt will impact your results. There are many conversation intelligence tools available. Some predesigned prompts display the resulting analysis in a consumable form like text, a visual (smiley and sad face icons), or a graph. Many of these tools also allow you to design your prompts. As my friend and colleague Chris Cognetta says,

"Get ready to add prompt engineering to your skill set because it will become part of your daily lives before you know it."

Chapter 9:
The Path to Insights

So here you are! You learned the importance of a conversation intelligence mindset and that technology helps you analyze it in new and powerful ways. You did your due diligence, recruited stakeholders, and built a strategic plan. But how do you tell the story your conversation intelligence data is telling you?

When working with conversation intelligence data, you deal with a lot of information. It can be challenging to understand the big picture of what's going on in your conversations without some visual representation. That's where data visualization comes in - it allows you to see patterns and trends in your data that might not be immediately obvious otherwise.

Using data visualization to tell your conversation intelligence story, you can straightforwardly communicate your findings. You can create charts, graphs, and other visual aids that help you convey your message impactfully. This can be especially important if you're trying to communicate your findings to stakeholders or other decision-makers who may not be as familiar with the data as you are.

The human brain can process visual information up to 60,000 times faster than text-based information. So, consider data visualization an essential part of your conversation intelligence success plan.

Visual Dashboards Benefits

If you use CRM and Business Intelligence (BI) tools, you already know many of the benefits of seeing data represented in a consumable and visual element like a chart or graph. Conversation intelligence dashboards and visualizations also offer essential benefits by providing a comprehensive view of customer conversations across multiple communication channels.

By presenting conversation intelligence information in an easily digestible format, you ensure that the right insights get to the right stakeholders and have the insights they need to make informed business decisions.

If you are in one of the highly regulated industries we discussed earlier in the book (like finance and healthcare), then the ability to help you meet regulatory requirements in real-time to ensure compliance is imperative.

Conversation intelligence dashboards and visualizations can help you visualize trends and patterns in customer behavior and preferences that are difficult to perceive otherwise, especially across functions. This is particularly true if your company adopts a RevOps strategy requiring visualizations representing conversations throughout the customer journey.

Rev Ops, short for Revenue Operations, is a business strategy that aims to align sales, marketing, and customer success teams to optimize revenue growth and customer loyalty. It involves streamlining processes and workflows, leveraging technology, and fostering communication and collaboration between teams to ensure that every aspect of the customer journey is optimized for revenue growth and customer-focused outcomes. Conversation intelligence is a crucial component of this strategy.

Here are some specific benefits of using conversation intelligence data visualization:

- **Increased Data Accessibility:** Visual dashboards and representations of conversation intelligence data can make data more accessible to key stakeholders across your company. We already discussed stakeholders who may be concerned about the impact and risks of using conversation intelligence. Here is your opportunity to increase transparency in those areas of concern. Data accessibility improves collaboration and better decision-making.

- **Enhanced Data Analysis:** Visual dashboards can provide a more comprehensive view of the data, allowing you to identify patterns and trends that may not be apparent in raw data. It can lead to more accurate and insightful analysis, which can help you make better decisions for your business. If you organize under the Rev Ops structure I discussed earlier, you can better recognize conversation trends across channels and functions across your business.

- **Improved Communication:** Visual dashboards and representations of conversation intelligence data can help you improve communication between different departments and stakeholders across your company, even if you are not in a Rev Ops -structured organization. Visualizing a challenge increases opportunities to discuss it and create solutions collaboratively.

- **Real-Time Monitoring:** In some cases, seeing trends or challenges happen real-time allows you to address issues as they arise. Responding to and mitigating crises is invaluable for highly regulated industries. However, adjusting marketing messages in response to negative customer sentiment can save your company money. Any scenario that involves discovering and addressing challenges in real-time improves perceptions or addresses customer challenges are worth the investment in visualization.

- **Better Resource Allocation:** You can also improve operational efficiency and reduce costs by identifying areas for resource allocation improvement. All that conversation intelligence comes from real conversations with people who spend time with customers and other internal stakeholders and partners. Visualizing how much time people spend accomplishing tasks and the resources that support them helps you make better resource allocation decisions.

- **Improved Decision-Making:** Although intuition is helpful in many situations, running a business solely based on intuition is risky. Visual dashboards and representations of conversation intelligence data can provide key stakeholders across your company with the information they need to make better business decisions based on accurate data. If you use voice-of-the-employee and voice-of-the-customer surveys, conversation intelligence can literally add voices to those insights.

- **Increased Accountability:** By providing key stakeholders across your company with access to conversation intelligence data, you can increase accountability and ensure everyone is working towards the same goals. The playbooks we discussed earlier as coaching tools in conversation intelligence can also serve as an accountability measurement. By visualizing deficiencies in using playbooks, we can help our teams follow them more closely.

- **Improved Forecasting:** You can improve your forecasting capabilities by analyzing customer conversations and identifying trends and patterns. Conversation intelligence data visualization helps you make more accurate predictions about future customer behavior and plan accordingly. Retail companies, in particular, can benefit from identifying upcoming changes in e-commerce behavior. Hospitality companies can also improve forecasting based on conversation intelligence about customer behavior changes. Conversations about increases in COVID-19 cases allowed many healthcare companies to forecast potential increases in emergency room visits and hospitalizations.

- **Enhanced Compliance:** Visual dashboards and representations of conversation intelligence data can help you ensure compliance with regulatory requirements and internal policies. Compliance reduces legal and reputational risk and improves customer trust and loyalty. Plus, you can avoid potential fines and penalties.

- **Increased Competitive:** Perhaps conversation intelligence and AI will not replace your lawyer or dentist. However, the lawyer or dentist leveraging conversation intelligence and AI will have a significant competitive advantage over those who do not. The same is true about you and your competitor. Understanding what conversation intelligence says about your customers and market will help you increase market share and improve your business outcomes above your competition's. Plus, you'll be able to differentiate yourself from your competitors and attract more customers. The ability to view this data visually and share it with your organization will keep you vigilant against competitors.

- **Improved Employee Performance:** Conversation intelligence dashboards and visualizations also help you identify areas where employees can improve their communication skills so you can provide targeted training and coaching to improve employee performance.

- **Improved Customer Retention:** Imagine visualizing the conversations that led to keeping or losing a customer. Would you change the dialogue to increase retention if you could? Would you adjust your engagement and improve your words to reiterate your value? Of course, you would. You can improve customer retention rates by analyzing customer conversations and identifying areas where you can improve the customer experience.

- **Increased Efficiency:** Just as dashboards can tell you if you are accountable to business objectives, they can also tell you where you are not efficiently meeting them. Visual dashboards and representations of conversation intelligence data can help you identify areas where you can streamline processes. They can help you determine where to reduce costs or do more in less time and with fewer resources.

- **Better Risk Management:** By analyzing customer conversations and identifying potential risks, you can develop strategies to mitigate these risks. Improving risk management leads to reduced legal and reputational risk and improved customer trust and loyalty. You'll be able to avoid potential issues and be better prepared for potential threats. When you incorporate alerts that leverage conversation intelligence insights to act, you also improve risk management.

- **Improved Employee Engagement:** You can increase employee engagement and motivation by providing employees with access to conversation intelligence data. Leaderboards and gamification are also effective ways to use dashboards to engage employees and motivate them to do their best work.

Creating visual dashboards and representations of conversation intelligence data can provide a range of benefits to your company. There are many vendor solutions available for data representation and visualization that I will not discuss in this book. However, I want to discuss some best practices that will increase your chances of getting conversation intelligence insights to the right people in the right ways.

Best Practices for Presentation

- **Keep It Simple:** When presenting conversation intelligence analysis, it is important to keep the data simple and easy to understand. Use clear and concise language and avoid technical jargon.

- **Focus On Key Insights:** Focus on the insights most relevant to the audience. Avoid overwhelming the audience with too much data.

- **Provide Context:** Explain the methodology used to collect and analyze the data and any limitations or caveats.

- **Customize for the Audience:** Tailor the data and visuals to the specific needs and interests of the audience to ensure that the presentation is relevant and engaging.

- **Use Multiple Data Sources:** Providing a more comprehensive view of the data can help you identify patterns and trends that may not be apparent in a single data source.

- **Provide Actionable Insights:** Actionable insights that the audience can implement are the most meaningful. Value and relevance to the audience should be your priority.

- **Use A Storytelling Approach:** Use storytelling to help engage the audience and make the data more memorable. Use anecdotes and examples to illustrate key points and make the data more relatable.

- **Use Benchmarking:** Compare the data to industry standards and competitors. It will help provide context for the data and identify areas where the company can improve.

- **Include Recommendations:** Include recommendations for action based on the data. It is not apparent to people how data will help them make business decisions and lead to improved business outcomes. You can use the previous section of this chapter to make recommendations for how this data can help.

- **Ensure Data Quality:** Ensure that the data is accurate and reliable. Use quality control measures to verify the data and ensure that it is representative of the target population. Lack of data accuracy can cause people to distrust similar data in the future.

- **Provide Training:** Train key stakeholders across the company to ensure they understand how to use the data effectively. There is a lot of power in conversation intelligence data, and even more when people are empowered to make sense of it independently.

- **Use A Consistent Format:** A consistent format for the data and visuals makes the presentation easy to follow and understand. It makes it easier for the audience to compare different data sets.

- **Be Transparent:** Be transparent about the data sources, methodology, and limitations. You will make mistakes as you evaluate how to present information to multiple audiences best. Transparency will help you build trust with the audience and ensure the credibility of the presentation.

- **Share Success Stories:** Share success stories and examples of how the data has been used to drive positive business outcomes. It will help inspire the audience and demonstrate the value of the data.

- **Follow Up:** Follow up with the audience to ensure they understand the data and can apply it effectively. Don't just post a report somewhere and hope for the best. Following up will help ensure that the presentation has a lasting impact and accelerates value outcomes.

The best practices I shared for presenting conversation intelligence analysis are a good start. I encourage you to partner with your company's analyst and data scientist to better understand how to share conversation intelligence data. The goal is always to find better and more innovative ways to give people answers to the questions they are asking. Maybe even answer questions they didn't know they had.

Integration For Success

Having suggested several times that there is a lot of value to analyzing conversations across many channels and functions, I don't want to overlook discussing integration to other technologies and enterprise business applications and tools. By combining conversation intelligence with other data sources, integration will give you a more comprehensive view of your customers, behavior, and preferences. Integration can help you communicate with your customers more effectively and tailor your approach to their needs throughout the customer journey.

Integrating conversation intelligence with other applications and tools can also help you operate more efficiently and effectively. You can reduce manual effort and improve productivity by automating tasks and processes that leverage conversation intelligence.

Here are some considerations for integration with the most common platforms that leverage conversation intelligence:

Customer Relationship Management (CRM) Systems

Integration with CRM systems can provide a comprehensive view of customer interactions and enable personalized communication with customers. Currently, the top two CRM vendors in the market include native integration to conversation intelligence tools that gather conversation data from email and unified communications. Conversation intelligence in these CRM systems helps users with the transcription and summarization of conversations, making it easy to understand the context of these conversations. These summaries can then be recorded in the history of an account, contact, opportunity, or service record.

Cons:

If your CRM system does not have native conversation intelligence capabilities, integration can be complex and require significant resources. If you are using other data sources within the CRM system to conduct conversation intelligence analysis, consider that the quality of the data in the CRM system can impact the accuracy of the conversation intelligence analysis.

Marketing Automation Platforms

If marketing automation is part of the CRM systems I just discussed, they, too, can leverage conversation intelligence analysis. Marketing automation platforms can provide conversation insights into customer behavior that help you improve targeted marketing campaigns. Conversation intelligence can help identify areas where marketing automation can be improved, leading to increased efficiency and effectiveness.

Cons:

As with CRM integration, if conversation intelligence is not native, it can be complex and require significant resources. The quality of the data in the marketing automation platform can also impact the conversation intelligence analysis's accuracy.

Other Integrations

Other potential integrations that could add potential business value and insights include the following:

- **Customer Experience Management (CEM) Systems:** Can provide insights into the customer experience and help identify ways to make the experience more functional, effective, accessible, and emotionally relevant.

- **Business Intelligence (BI) Platforms:** Integration with BI platforms can provide a comprehensive view of business performance and enable data-driven decision-making.

- **Natural Language Processing (Nlp) Tools:** NLP tool integration can improve the accuracy of the conversation intelligence analysis and enable more advanced study.

- **Chatbots and Virtual Assistants:** Integration with chatbots and virtual assistants can enable real-time customer communication and provide insights into customer behavior.

- **Voice Recognition Software:** Like chatbots and virtual assistants, the greatest integration benefit is enabling real-time analysis of customer conversations and providing insights into customer behavior. The same is true about integration with **Social Media Monitoring Tools.**

Also true about data integration, in general, is that the quality of the data can impact the accuracy of the conversation intelligence analysis.

Conclusion & Next Steps

Conversation intelligence has many benefits, but they do not replace the need for careful strategic planning. Using methodologies like human-centered design thinking, you can gather stakeholders to understand the impact of conversation intelligence in their organizations and roles.

- Identify the business needs and goals you want to achieve with conversation intelligence.

- Evaluate the different conversation intelligence tools available on the market. Select a platform that aligns with your needs and goals.

- Design effective prompts that capture your conversations' most relevant and actionable insights.

- Test and refine your prompts over time based on feedback and insights from your conversations.

- Collaborate with stakeholders from across the

organization to ensure your conversation intelligence strategy aligns with business goals and objectives.

Use the insights you gain from conversation intelligence to improve your business processes and achieve better results.

Chapter 10:
The Future is Now

Conversation intelligence is a relatively new technology rapidly evolving and transforming how you interact with your customers. You could compare it to other quickly evolving technologies, such as artificial intelligence and machine learning.

Like conversation intelligence, artificial intelligence and machine learning started as relatively new and untested technologies. However, over time, they have become more sophisticated and powerful. We now use them in various applications, from self-driving cars and personal assistants like Siri and Alexa.

Similarly, conversation intelligence is evolving rapidly and will likely become even more powerful and sophisticated. We expect to see even more benefits as more companies adopt and integrate this technology with other enterprise business applications and tools.

Despite its impressive functionality and increasing adoption, conversational intelligence is still in its infancy. Still, it has the potential to transform how companies interact with their customers in every aspect of business. As this technology continues to evolve and mature, we can expect to see even more exciting developments in the future.

Remaining Challenges

Having said that, it is expected that the technology we discussed in detail throughout the book (the engine driving conversation intelligence) will be the first to achieve next-generation innovation and evolution:

- **AI & ML**: One area where conversation intelligence will likely continue to grow is using artificial intelligence (AI) and machine learning (ML) algorithms. These technologies can enable more advanced analysis of customer conversations, including more advanced types of sentiment analysis like the ones we discussed in previous chapters, topic modeling, and predictive analytics.

- **NLP**: Another area where conversation intelligence will likely continue to grow is the use of natural language processing (NLP) and voice recognition technologies. These technologies can enable more accurate and efficient analysis of customer conversations. They can allow real-time communication with customers through chatbots and virtual assistants.

However, conversation intelligence areas for improvement remain. Technologists are still working on specific areas of challenge, such as:

- **Accuracy**: One area is in the accuracy of the analysis, particularly in cases where there is noise or ambiguity in the data.

- **Integration**: Another area is integrating other technologies and enterprise business applications and tools, which can be complex and require significant resources.

Despite these challenges, the outlook for conversation intelligence is hopeful. According to a recent report by MarketsandMarkets, the conversation intelligence market is expected to grow from $4.8 billion in 2020 to $13.8 billion by 2025 at a compound annual growth rate (CAGR) of 23.9% (MarketsandMarkets, 2020). This growth is driven by increasing demand for customer insights and the need for more efficient and effective communication with customers.

Focusing On Ethics

In addition, there is a growing focus on the ethical use of conversation intelligence, particularly in data privacy and bias areas. Companies increasingly recognize the importance of responsible use of customer data and ensuring that conversation intelligence is used fairly and transparently.

Multichannel

Companies increasingly use multiple channels to communicate with their customers, including email, chat, social media, and phone. Conversation intelligence can enable companies to analyze customer interactions across these channels and provide a more seamless and personalized communication experience. However, customers use many channels that are still not integrated to conversation intelligence and often use a combination of channels within a single conversation.

Real–Time & Augmented Analytics

Real-time analytics can enable companies to identify and respond to customer issues and opportunities in real time, improving the customer experience and driving business outcomes. This technology is particularly important in industries such as healthcare, where real-time communication can be critical to patient outcomes. However, presenting conversation intelligence insight in real-time during conversations, and expecting people to change their behavior or correct their course of action, requires additional study. Augmented analytics, in particular, can enable companies to automate customer conversation analysis and provide real-time insights without human intervention. While it has seen some success in the call center space, augmented analytics struggles to deliver effective service when deductive reasoning is required.

Standardization

Another challenge is the need for more standardized data formats and integration protocols. Standardization allows easier integration with other technologies and enterprise business applications and tools. It can improve the accuracy and efficiency of the analysis. The development of more standardized integration protocols and data formats that will enhance the quality and consistency of the data is in progress.

Transparent and Ethical Use

As this technology becomes more widespread, companies need to be transparent about how they use customer data and ensure that their use of conversation intelligence is fair and ethical. As of this book's writing, no regulatory agency is creating common standards of practice specific to conversation intelligence. Therefore, accountability for the ethical use of conversation intelligence falls on every company.

Conclusion & Next Steps

Congratulations! You've made it to the end, which in many ways, is the beginning of your journey. By now, you should have a solid understanding of what conversation intelligence is, how it works, and how it can benefit your business. But if you're still feeling a little unsure or hesitant about using conversation intelligence, don't worry - you're not alone.

The truth is, implementing new technology or processes can be intimidating, especially if you're not familiar with the tools or techniques involved. But the good news is that conversation intelligence already has an impressive success record with many companies you know. And with the right approach and mindset, you can overcome any challenges or obstacles that come your way. Hopefully, I have gained your trust in supporting your success.

As we conclude the book, I want to share a few final tips to help you feel more confident and empowered as you begin your conversation intelligence journey:

- **Start small:** You don't have to implement conversation intelligence across your entire organization all at once. Start with a small pilot project or team, and gradually expand as you gain more experience and confidence.

- **Focus on the benefits:** Rather than getting bogged down in the technical details or potential challenges of conversation intelligence, focus on the benefits it can bring to your business. By improving your customer interactions and employee performance, you can achieve better results and drive growth for your organization.

- **Embrace a growth mindset:** Remember that implementing conversation intelligence is a learning process, and you will inevitably encounter challenges or setbacks. But by embracing a growth mindset and viewing these challenges as opportunities for growth and improvement, you can overcome them and achieve even greater success.

- **Collaborate with others:** Don't be afraid to collaborate with stakeholders across your organization, such as sales, marketing, customer service, and HR. By working together and sharing insights and feedback, you can develop a more effective conversation intelligence strategy that aligns with your business goals and objectives.

In conclusion, conversation intelligence can help you achieve extraordinary results for your business. So go forth confidently, and remember you have the knowledge and skills to make conversation intelligence work for you!

Oh, and as for Sarah and the team at Sweet Treats, while the names have been changed, their success is real, and a great example of the success you and your company can also achieve.

References

Chapter 1

Bain & Company. (2019). The Elements of Value®: Measuring—and delivering—what consumers really want. https://www.bain.com/contentassets/7a5b9f9c8f5a4d5b8d5d5f1b8e5e5f5f/the-elements-of-value.pdf

BrightLocal. (2020). Local consumer review survey 2020. https://www.brightlocal.com/research/local-consumer-review-survey/

Harvard Business Review. (2011). The short life of online sales leads. https://hbr.org/2011/03/the-short-life-of-online-sales

HubSpot. (2018). The ultimate list of sales statistics for 2018. https://blog.hubspot.com/sales/sales-statistics

Journal of Business and Psychology. (2017). The role of dialogue in creativity and commitment in organizations. https://link.springer.com/article/10.1007/s10869-016-9473-3

Journal of Business Research. (2019). Customer engagement and customer experience: What they are and how they affect customer behavior. https://www.sciencedirect.com/science/article/abs/pii/S0148296319305709

Journal of Marketing Research. (2016). The impact of customer participation on customer satisfaction and intended loyalty in B2B relationships. https://journals.sagepub.com/doi/abs/10.1177/0022243716659082

McKinsey & Company. (2016). The CEO guide to customer experience. https://www.mckinsey.com/business-functions/operations/our-insights/the-ceo-guide-to-customer-experience

HubSpot. (2019). The state of customer service in 2019. https://blog.hubspot.com/service/customer-service-statistics

Chapter 2

Ambady, N., & Rosenthal, R. (1992). Thin slices of expressive behavior as predictors of interpersonal consequences: A meta-analysis. Psychological Bulletin, 111(2), 256–274.

Cherry, K. (2021). Selective Attention. Verywell Mind. https://www.verywellmind.com/what-is-selective-attention-2795026

Cowan, N. (1995). Attention and Memory: An Integrated Framework. Oxford University Press.

Dweck, C. S. (2006). Mindset: The new psychology of success. Random House.

Fiske, S. T., & Taylor, S. E. (2013). Social cognition: From brains to culture. Sage Publications.

Glaser, J. E. (2019). Conversational Intelligence: How Great Leaders Build Trust & Get Extraordinary Results. Routledge.

Goleman, D. (1995). Emotional intelligence. Bantam Books.

Heath, C., & Heath, D. (2010). Switch: How to change things when change is hard. Crown Business.

Hsee, C. K., Hatfield, E., Carlson, J. G., & Chemtob, C. (1990). The effect of power on susceptibility to emotional contagion. Cognition and Emotion, 4(4), 327-340.

Kegan, R., & Lahey, L. L. (2009). Immunity to change: How to overcome it and unlock potential in yourself and your organization. Harvard Business Press.

Levine, M. (2013). The positive psychology of Buddhism and yoga: Paths to a mature happiness. Routledge.

Mayer, J. D., Salovey, P., & Caruso, D. R. (2004). Emotional intelligence: Theory, findings, and implications. Psychological Inquiry, 15(3), 197-215.

Mehrabian, A., & Ferris, S. R. (1967). Inference of attitudes from nonverbal communication in two channels. Journal of consulting psychology, 31(3), 248.

Schein, E. H. (2010). Organizational culture and leadership. John Wiley & Sons.

Sinek, S. (2011). Start with why: How great leaders inspire everyone to take action. Penguin.

Tannen, D. (1991). You just don't understand: Women and men in conversation. Ballantine Books.

Thibodeaux, M. S. (2019). The introverted leader: Building on your quiet strength. Berrett-Koehler Publishers.

Vogt, K. (2016). Collaboration begins with you: Be a silo buster. Berrett-Koehler Publishers.

Watzlawick, P., Beavin-Bavelas, J., & Jackson, D. D. (2011). Pragmatics of human communication: A study of interactional patterns, pathologies, and paradoxes. WW Norton & Company.

Weger Jr, H., Castle Bell, G., Minei, E. M., & Robinson, M. C. (2014). The relative effectiveness of active listening in initial interactions. International Journal of Listening, 28(1), 13-31.

Weick, K. E., & Sutcliffe, K. M. (2007). Managing the unexpected: Resilient performance in an age of uncertainty. John Wiley & Sons.

Zak, P. J. (2017). The neuroscience of trust. Harvard Business Review, 95(1), 84-90.

Chapter 3

Aberdeen Group. (2019). The impact of account-based marketing with conversation intelligence. Retrieved from https://www.aberdeen.com/research-report/the-impact-of-account-based-marketing-with-conversation-intelligence/

McKinsey & Company. (2020). Agile in the enterprise. Retrieved from https://www.mckinsey.com/business-functions/mckinsey-digital/our-insights/agile-in-the-enterprise

Morgan, J. (2021). How conversation intelligence can help sales teams. Harvard Business Review. Retrieved from https://hbr.org/2021/07/how-conversation-intelligence-can-help-sales-teams

Rogers, M. (2019). The value of inbound marketing with conversation intelligence. Retrieved from https://www.hubspot.com/inbound-marketing

Six Sigma Daily. (2021). Six Sigma statistics: How to measure success. Retrieved from https://www.sixsigmadaily.com/six-sigma-statistics-how-to-measure-success/

Ward, J., & Peppard, J. (2016). The strategic role of IT: An empirical study of its effects on innovation. Journal of Strategic Information Systems, 25(1), 1-20. doi:10.1016/j.jsis.2015.11.001

Bock, A. J. (2017). Lean Six Sigma in healthcare: A powerful tool for improving quality and safety. Journal of Healthcare Risk Management, 36(2), 6-13. doi:10.1002/jhrm.21186

Gibson, J. L., Ivancevich, J. M., & Donnelly, J. H. (2012). Organizations: Behavior, structure, processes. New York, NY: McGraw-Hill.

Kotler, P., & Keller, K. L. (2016). Marketing management (15th ed.). Upper Saddle River, NJ: Pearson.

Mintzberg, H. (1994). The rise and fall of strategic planning. New York, NY: Free Press.

Porter, M. E. (1996). What is strategy? Harvard Business Review, 74(6), 61-78.

Senge, P. M. (1990). The fifth discipline: The art and practice of the learning organization. New York, NY: Doubleday/ Currency.

Tidd, J., & Bessant, J. (2018). Managing innovation: Integrating technological, market and organizational change (6th ed.). Chichester, UK: Wiley.

Womack, J. P., Jones, D. T., & Roos, D. (1990). The machine that changed the world: The story of lean production. New York, NY: Rawson Associates.

Chapter 4

Boles, J. S., Madupalli, R., & Kahle, L. R. (2018). A conceptual framework of key account management effectiveness. Journal of Business Research, 88, 388-398.

Cusumano, M. A., & Selby, R. W. (2018). Microsoft: Cloud computing and the cloud platform wars. Communications of the ACM, 61(4), 34-43.

Dawes, P. L., Massey, G. R., & Soutar, G. N. (2018). Key account management and its impact on customer-oriented behavior and sales growth. Journal of Personal Selling & Sales Management, 38(4), 372-389.

Dixon, M., Adamson, B., & Toman, N. (2012). The challenger sale: Taking control of the customer conversation. Penguin.

Foss, N. J., & Saebi, T. (2017). Fifteen years of research on business model innovation: How far have we come, and where should we go?. Journal of Management, 43(1), 200-227.

Kapoor, R., & Pandey, S. (2018). Relationship management framework for a sustainable supply chain. Journal of Cleaner Production, 172, 2587-2596.

Kumar, V., & Reinartz, W. (2018). Customer relationship management: Concept, strategy, and tools. Springer.

Miller, H., & Heiman, S. (2017). The new strategic selling: The unique sales system proven successful by the world's best companies. Grand Central Publishing.

Rackham, N. (2010). Spin selling. McGraw-Hill Education.

Reilly, T., & Forde, T. (2018). Target account selling: The ultimate sales tool. John Wiley & Sons.

Vengoechea, R., & Selden, L. (2018). Value-based selling: How to sell more profitably, confidently, and professionally by getting customers to understand and buy your value proposition. John Wiley & Sons.

Chapter 5

Harvard Business Review. (2019). The new science of customer emotions. Retrieved from https://hbr.org/2015/11/the-new-science-of-customer-emotions

Forrester Research. (2018). The top 10 customer service trends for 2019. Retrieved from https://go.forrester.com/blogs/the-top-10-customer-service-trends-for-2019/

Gartner. (2019). Market intelligence. Retrieved from https://www.gartner.com/en/information-technology/glossary/market-intelligence

Huang, R., & Sarigöllü, E. (2014). How brand awareness relates to market outcome, brand equity, and the marketing mix. In Fashion Branding and Consumer Behaviors (pp. 113-132). Springer, New York, NY.

Kaplan, R. S., & Norton, D. P. (2001). The strategy-focused organization: How balanced scorecard companies thrive in the new business environment. Harvard Business Press.

Keller, K. L. (2008). Strategic brand management: Building, measuring, and managing brand equity. Pearson Education India.

Kotler, P., & Armstrong, G. (2016). Principles of marketing. Pearson Education Limited.

Morgan, N. A., Vorhies, D. W., & Mason, C. H. (2009). Market orientation, marketing capabilities, and firm performance. Strategic Management Journal, 30(8), 909-920.

Porter, M. E. (1985). Competitive advantage: Creating and sustaining superior performance. Free Press.

Srinivasan, S., & Hanssens, D. M. (2009). Marketing and firm value: Metrics, methods, findings, and future directions. Journal of Marketing Research, 46(3), 293-312.

Wang, Y., Lo, H. P., & Yang, Y. (2004). An integrated framework for customer value and customer-relationship-management performance: A customer-based perspective from China. Managing Service Quality: An International Journal, 14(2/3), 169-182.

Chapter 6

Lerner, J. S., Li, Y., Valdesolo, P., & Kassam, K. S. (2015). Emotion and decision making. Annual review of psychology, 66, 799-823.

Kim, S., & Kwon, Y. (2018). Sentiment analysis for decision making in social media. Journal of Business Research, 89, 1-10.

Goudarzi, S., & Zarei, B. (2019). The role of sentiment analysis in decision making. Journal of Decision Systems, 28(1), 1-14.

Karniol, R., Grosz, E., & Schorr, I. (2003). Empathy as a predictor of inflammation. Journal of Social and Clinical Psychology, 22(6), 709-728. doi: 10.1521/jscp.22.6.709.22931

Neff, K. D., & Germer, C. K. (2013). A pilot study and randomized controlled trial of the mindful self-compassion program. Journal of Clinical Psychology, 69(1), 28-44. doi: 10.1002/jclp.21923

Piferi, R. L., Lawler, K. A., & Jobe, R. L. (2006). Giving to others and the association between stress and mortality. American Journal of Public Health, 96(11), 2245-2248. doi: 10.2105/AJPH.2005.072595

University of Michigan. (2014). Empathy makes us more social. Retrieved from https://news.umich.edu/empathy-makes-us-more-social/

University of Arizona. (2013). Compassionate acts and inflammation. Retrieved from https://uanews.arizona.edu/story/compassionate-acts-and-inflammation

Chapter 7

World Economic Forum. (2020). Global AI Action Alliance: Guidance on AI Ethics and Governance. Retrieved from https://www.weforum.org/reports/global-ai-action-alliance-guidance-on-ai-ethics-and-governance

European Commission. (2019). Ethics Guidelines for Trustworthy AI. Retrieved from https://ec.europa.eu/digital-single-market/en/news/ethics-guidelines-trustworthy-ai

Privacy Concerns:

Bhatia, A., & Jain, A. (2018). Privacy preserving techniques for big data analytics: A survey. Journal of Big Data, 5(1), 1-35. doi: 10.1186/s40537-018-0146-0

Gürses, S., Troncoso, C., & Diaz, C. (2011). Engineering privacy by design. Computers, Privacy and Data Protection, 11(1), 1-34. Retrieved from https://papers.ssrn.com/sol3/papers.cfm?abstract_id=1811002

Hildebrandt, M., & Gutwirth, S. (2016). Profiling and big data: A European approach. Computer Law & Security Review, 32(2), 256-271. doi: 10.1016/j.clsr.2015.12.006

Kshetri, N., & Voas, J. (2018). Blockchain-enabled secure sharing of electronic health records: A review. Decision Support Systems, 113, 1-14. doi: 10.1016/j.dss.2018.06.005

Solove, D. J. (2013). Privacy self-management and the consent dilemma. Harvard Law Review, 126(7), 1880-1903. doi: 10.2139/ssrn.2187873

References

Accuracy Concerns:

Davenport, T. H., & Patil, D. J. (2012). Data scientist: The sexiest job of the 21st century. Harvard Business Review, 90(10), 70-76. Retrieved from https://hbr.org/2012/10/data-scientist-the-sexiest-job-of-the-21st-century

Domingos, P. (2015). The master algorithm: How the quest for the ultimate learning machine will remake our world. Basic Books.

Provost, F., & Fawcett, T. (2013). Data science and its relationship to big data and data-driven decision making. Big Data, 1(1), 51-59. doi: 10.1089/big.2013.1508

Bias Concerns:

Barocas, S., & Selbst, A. D. (2016). Big data's disparate impact. California Law Review, 104(3), 671-732. doi: 10.2139/ssrn.2477899

Buolamwini, J., & Gebru, T. (2018). Gender shades: Intersectional accuracy disparities in commercial gender classification. Proceedings of the Conference on Fairness, Accountability, and Transparency, 77-91. doi: 10.1145/3178876.3186150

Caliskan, A., Bryson, J. J., & Narayanan, A. (2017). Semantics derived automatically from language corpora contain human-like biases. Science, 356(6334), 183-186. doi: 10.1126/science.aal4230

Friedler, S. A., Scheidegger, C. E., Venkatasubramanian, S., Choudhary, S., Hamilton, E. P., & Roth, D. (2019). A comparative study of fairness-enhancing interventions in machine learning. Proceedings of the Conference on Fairness, Accountability, and Transparency, 329-338. doi: 10.1145/3287560.3287589

Kleinberg, J., Ludwig, J., Mullainathan, S., & Sunstein, C. R. (2018). Discrimination in the age of algorithms. Journal of Legal Analysis, 10(1), 113-174. doi: 10.1093/jla/lax012

Legal Concerns:

Bellovin, S. M. (2018). The legal implications of big data. Communications of the ACM, 61(4), 18-20. doi: 10.1145/3180492

Hartzog, W. (2018). Privacy's blueprint: The battle to control the design of new technologies. Harvard University Press.

Ohm, P. (2010). Broken promises of privacy: Responding to the surprising failure of anonymization. UCLA Law Review, 57(6), 1701-1777. Retrieved from https://papers.ssrn.com/sol3/papers.cfm?abstract_id=1450006

Schwartz, P. M. (2015). Data breach notification laws. Journal of Information Privacy and Security, 11(3), 94-108. doi: 10.1080/15536548.2015.1083264

Solove, D. J. (2013). Privacy self-management and the consent dilemma. Harvard Law Review, 126(7), 1880-1903. doi: 10.2139/ssrn.2187873

Swire, P. P. (2017). The EU general data protection regulation: What the US healthcare sector needs to know. Journal of AHIMA, 88(6), 26-31. Retrieved from https://journal.ahima.org/the-eu-general-data-protection-regulation-what-the-us-healthcare-sector-needs-to-know/

Transparency Concerns:

Acquisti, A., & Grossklags, J. (2005). Privacy and rationality in individual decision making. IEEE Security & Privacy, 3(1), 26-33. doi: 10.1109/MSP.2005.7

Balebako, R., Liu, F., & Marshini, S. (2015). The impact of personalization on Airbnb trust. Proceedings of the Conference on Human-Computer Interaction, 337-346. doi: 10.1145/2702123.2702158

Cranor, L. F. (2012). Necessary but not sufficient: Standardized mechanisms for privacy notice and choice. IEEE Security & Privacy, 10(2), 68-73. doi: 10.1109/MSP.2012.44

Jensen, C., Potts, C., & Jensen, C. D. (2005). Privacy practices of internet users: Self-reports versus observed behavior. International Journal of Human-Computer Studies, 63(1-2), 203-227. doi: 10.1016/j.ijhcs.2005.04.007

Xu, H., Teo, H. H., Tan, B. C. Y., & Agarwal, R. (2009). The role of push-pull technology in privacy calculus: The case of location-based services. Journal of Management Information

Systems, 26(3), 135-174. doi: 10.2753/MIS0742-1222260305

Security Concerns:

Anderson, R. (2008). Security engineering: A guide to building dependable distributed systems. John Wiley & Sons.

Bishop, M. (2003). Computer security: Art and science. Addison-Wesley Professional.

Greenberg, A. (2014). This machine kills secrets: How WikiLeakers, cypherpunks, and hacktivists aim to free the world's information. Penguin.

Ross, J. W., & Blumenstein, M. (2017). Cybersecurity and privacy: An overview. Journal of Business Logistics, 38(1), 1-4. doi: 10.1111/jbl.12157

Schneier, B. (2015). Data and Goliath: The hidden battles to collect your data and control your world. WW Norton & Company.

Ethical Concerns:

Floridi, L. (2016). The ethics of information. Oxford University Press.

Johnson, D. G., & Nissenbaum, H. (1995). Computer systems: Moral entities but not moral agents. Ethics and Information Technology, 1(1), 31-39. doi: 10.1007/BF01256270

Tavani, H. T., & Moor, J. H. (2001). Privacy protection, control of information, and privacy-enhancing technologies. Computers and Society, 31(3), 6-11. doi: 10.1145/384017.384019

Van den Hoven, J., & Weckert, J. (2008). Information technology and moral philosophy. Cambridge University Press.

Zimmer, M. (2010). Privacy and publicity in the context of big data. Ethics and Information Technology, 12(2), 119-132. doi: 10.1007/s10676-010-9227-5

Chapter 8

Fleming, J. S., & Levie, W. H. (2018). Conversation intelligence: How great leaders build trust and get extraordinary results. Routledge.

Liedtka, J., & Ogilvie, T. (2011). Designing for growth: A design thinking toolkit for managers. Columbia University Press.

Peter, J. P., & Donnelly, J. H. (2013). A preface to marketing management. McGraw-Hill Education.

Schumann, J. H., & Thorson, E. (2016). Benchmarking in advertising research. In Advertising theory (pp. 397-413). Routledge.

Smith, R. D., & Blazek, R. (2019). Conversation intelligence: How to build trust, improve engagement, and drive results. Wiley.

Planning:

Bryson, J. M. (2018). Strategic planning for public and nonprofit organizations: A guide to strengthening and sustaining organizational achievement. John Wiley & Sons.

Lorange, P. (2017). Strategic planning: A pragmatic guide. World Scientific.

Mintzberg, H. (1994). The rise and fall of strategic planning: Reconceiving roles for planning, plans, planners. Simon and Schuster.

Communication:

DeVito, J. A. (2016). The interpersonal communication book. Pearson.

Gudykunst, W. B., & Kim, Y. Y. (2017). Communicating with strangers: An approach to intercultural communication. Routledge.

O'Hair, D., Wiemann, M., Mullin, D. A., & Teven, J. J. (2015). Real communication: An introduction. Bedford/St. Martin's.

Emotional intelligence:

Goleman, D. (1995). Emotional intelligence: Why it can matter more than IQ. Bantam.

Mayer, J. D., & Salovey, P. (1997). What is emotional intelligence? In P. Salovey & D. Sluyter (Eds.), Emotional development and emotional intelligence: Educational implications (pp. 3-31). Basic Books.

Salovey, P., & Mayer, J. D. (1990). Emotional intelligence. Imagination, cognition and personality, 9(3), 185-211.

Active listening:

Brownell, J. (2015). Active listening. In The international encyclopedia of interpersonal communication (pp. 1-5). John Wiley & Sons.

Burley-Allen, M. (1995). Listening: The forgotten skill: A self-teaching guide. John Wiley & Sons.

Miller, K. (2015). Organizational communication: Approaches and processes. Cengage Learning.

Chapter 9

Tufte, E. R. (2001). The cognitive style of PowerPoint: Pitching out corrupts within. Cheshire, CT: Graphics Press.

Few, S. (2013). Show me the numbers: Designing tables and graphs to enlighten. Analytics Press.

Kosslyn, S. M. (2006). Graph design for the eye and mind. Oxford University Press.

Additional Resources

Conversation Intelligence:

Fleming, J. S., & Levie, W. H. (2018). Conversation intelligence: How great leaders build trust and get extraordinary results. Routledge.

Liedtka, J., & Ogilvie, T. (2011). Designing for growth: A design thinking toolkit for managers. Columbia University Press.

Schumann, J. H., & Thorson, E. (2016). Benchmarking in advertising research. In Advertising theory (pp. 397-413). Routledge.

Smith, R. D., & Blazek, R. (2019). Conversation intelligence: How to build trust, improve engagement, and drive results. Wiley.

AI in Business:

Brynjolfsson, E., & McAfee, A. (2014). The second machine age: Work, progress, and prosperity in a time of brilliant technologies. WW Norton & Company.

Brynjolfsson, E., & McAfee, A. (2017). The business of artificial intelligence. Harvard Business Review, 95(1), 53-62.

Kiron, D., Prentice, P. K., & Ferguson, R. B. (2018). The future of jobs: How artificial intelligence, robotics, and automation are transforming the workplace. MIT Sloan Management Review, 59(4), 1-10.

Davenport, T. H., & Ronanki, R. (2018). Artificial intelligence for the real world. Harvard Business Review, 96(1), 108-116.

Davenport, T. H., & Kalakota, R. (2019). The potential for artificial intelligence in healthcare. Future Healthcare Journal, 6(2), 94-98.

Natural Language Processing for Business:

Boyd, R. L., & Wilson, D. C. (2019). Natural language processing in accounting, auditing and finance: A synthesis of the literature with a roadmap for future research. Journal of Information Systems, 33(3), 1-22.

Chen, H., Chiang, R. H., & Storey, V. C. (2012). Business intelligence and analytics: From big data to big impact. MIS Quarterly, 36(4), 1165-1188.

Choudhury, M. D., & Gadiraju, U. (2019). Natural language processing in business: A conceptual review and future research directions. Decision Support Systems, 120, 65-77.

Jurafsky, D., & Martin, J. H. (2020). Speech and language processing (3rd ed.). Pearson.

Kietzmann, J. H., Hermkens, K., McCarthy, I. P., & Silvestre, B. S. (2011). Social media? Get serious! Understanding the functional building blocks of social media. Business Horizons, 54(3), 241-251.

Liu, B. (2012). Sentiment analysis and opinion mining. Synthesis Lectures on Human Language Technologies, 5(1), 1-167.

Get the series!

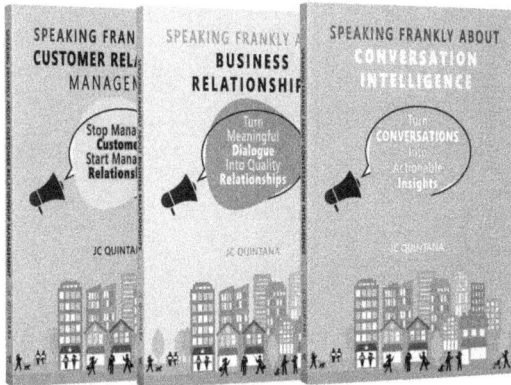

Speaking Frankly About Business Relationships

Business relationships need clear expectations. Do you know the most critical elements of business relationship dialogue and expectation management?

Speaking Frankly About Customer Relationships Management

Customer relationship management requires intentional goals for winning and keeping the right customer relationships. Do you know the ten most important conversations that lead to customer strategy and CRM success?

Speaking Frankly About Conversation Intelligence

There is untapped insight in your conversations. The ultimate competitive advantage is having the strategy and technology to gain intelligence from your conversations. Do you know where to start?

Available wherever books are sold.

Learn more at jcquintana.com